FINITE GROUPS '72

NORTH-HOLLAND
MATHEMATICS STUDIES **7**

Finite Groups '72

PROCEEDINGS OF THE THE GAINESVILLE CONFERENCE ON FINITE GROUPS, MARCH 23 - 24, 1972

Edited by
TERRENCE GAGEN
University of Sydney
Sydney, Australia

MARK P. HALE Jr.
ERNEST E. SHULT
University of Florida
Gainesville, Florida
U.S.A.

1973

NORTH-HOLLAND PUBLISHING COMPANY – AMSTERDAM • LONDON
AMERICAN ELSEVIER PUBLISHING COMPANY, INC. – NEW YORK

Library of Congress Catalog Card Number: 72 97697

ISBN North-Holland:

Series: 0 7204 2600 6

Volume: 0 7204 2607 3

ISBN American Elsevier: 0 444 10451 8

PUBLISHERS:

NORTH-HOLLAND PUBLISHING COMPANY – AMSTERDAM

NORTH-HOLLAND PUBLISHING COMPANY, LTD. – LONDON

SOLE DISTRIBUTORS FOR THE U.S.A. AND CANADA:

AMERICAN ELSEVIER PUBLISHING COMPANY, INC.

52 VANDERBILT AVENUE

NEW YORK, N.Y. 10017

PRINTED IN THE NETHERLANDS

RECEPTION PICTURES

TOP LEFT: Lyons, Wales, L. Scott, ________.

TOP RIGHT: Cassady, Walter.

MIDDLE LEFT: O'Nan, Schenkman, D. Higman.

MIDDLE RIGHT: Ito, Harada

BOTTOM LEFT: Chabot, Magliveras, _______, Phan, Gagen.
Srinivasan.

BOTTOM RIGHT: Teply, Kantor, Seitz.

TOP LEFT: Suzuki, Ito

TOP RIGHT: Gross, Aschbacher, Shult, Bednarek.

MIDDLE LEFT: O'Nan

MIDDLE RIGHT: University of Florida Renaissance Ensemble
(John Kitts, director)

BOTTOM LEFT: M. Hall, ________,

BOTTOM CENTER: Fong.

BOTTOM RIGHT: F. Hoffman, E. Ralston.

PREFACE

These proceedings consist of abstracts submitted by participants in the Gainesville Conference on Finite Groups held March 23 and 24, 1972, at the University of Florida. It is hoped that these papers will give some indication of the current state of the art of finite group theory. The editors wish to express their appreciation to the University of Florida Department of Mathematics for providing the financial support of this conference. We would also like to thank the North Holland Publishing Company for publishing these proceedings. Finally the editors take this opportunity to thank Mrs. Linda Carver and Mrs. Betty Holley for their valuable assistance in making arrangements for the conference and to thank Mrs. Vicki Shult for careful preparation of the manuscript for publication.

M.P.H.

T.G.

E.E.S.

TABLE OF CONTENTS

MINIMAL RESOLUTIONS

J. L. Alperin

The theory of blocks with cyclic defect groups is one of the high points of group theory; its extension to the non-cyclic case is the main problem of representation theory. Recently, the author and G. Janusz observed the following consequence of the theory: If G is a group with cyclic Sylow p-subgroup and F is a splitting field of characteristic p for G while

$$\cdots \to P_n \to P_{n-1} \to \quad \cdots\cdot \to P_0 \to F \to 0 \to \cdots$$

is a minimal projective resolution for F over F[G], then each projective module P_n is indecomposable. Moreover, the periodicity and the precise period are immediate.

A generalization of this result has been achieved for the case of dihedral Sylow 2-subgroups, in which the above "one-dimensional array" of indecomposable projective modules is replaced by a "two-dimensional array."

THEOREM 1. *If G is a group with dihedral Sylow 2-subgroups and F is a splitting field of characteristic two for G then there is a first quadrant double complex $\{P_{ij}\}$ of F[G]-modules with the following properties*:

1) *Each non-zero P_{ij} is an indecomposable projective module*;
2) *The associated single complex has an augmentation such that the augmented complex is the minimal projective resolution of F over F[G].*

In proving this one may assume that O(G) = 1. Moreover, by using the next result, one can then reduce to the case where G is

A_4 or has no proper normal subgroup of odd index.

THEOREM 2. *Let G be a normal subgroup of the group G_0, with G_0/G a solvable p'-group. Let F be a splitting field of characteristic p for G_0 and all its subgroups. Assume that the centralizer of a Sylow p-subgroup of G covers G_0/G. It follows that the restriction of $F[G_0]$-modules to F[G]-modules defines an isomorphism of the category of $F[G_0]$-modules, all of whose composition factors are in the principal p-block of G_0, onto the corresponding category for G.*

As for Theorem 1, we are now reduced to proving it when G is dihedral, PSL(2,q), PGL(2,q), q odd or A_7. The different congruences $q \equiv 1$ (modulo 4) and $q \equiv -1$ (modulo 4) lead to quite different arguments and so, in total, there are six separate cases to be dealt with. In each case detailed structure of all projective F[G]-modules in the principal 2-block is determined first and then the double complex is displayed. Other results on tensor products of irreducible modules and other blocks are obtained along the way.

A small start has been made in a similar study for groups with abelian Sylow 2-subgroups of order exceeding four. In particular, the Cartan matrix for p = 2 for $SL(2,2^n)$ has been determined and some results are known about the structure of projective modules in characteristic two for $SL(2,2^n)$.

University of Chicago

SYLOW 2-SUBGROUPS OF 2-RANK THREE

J. L. Alperin

The recent classifications of simple groups of 2-rank at most two and those of sectional 2-rank not exceeding four, with the consequent determination of simple groups of normal 2-rank at most two, lead us quite naturally to the next problem: If G is a fusion simple group of 2-rank three and normal 2-rank three, describe the Sylow 2-subgroups of G. This would be the first step in classifying all such groups.

If G is such a group and S is a Sylow 2-subgroup of G then S has a normal elementary abelian subgroup E of order eight but has no elementary abelian subgroup of larger order. The various possible structures for the automizer A of E lead to a number of cases, each to be dealt with separately. The first occurs when A is the entire automorphism group of E and then S can be completely determined. In fact, we have a more general result.

THEOREM 1. *Let* E *be a normal elementary abelian subgroup of order eight in a group* G *and assume the following conditions*:

a) G *has no normal non-identity subgroup of odd order*;
b) *The involutions of* C *are exactly the non-identity elements of* E;
c) *If* C *is the centralizer of* E *then* G/C *is isomorphic with the automorphism group of* E.

It follows that there is a positive integer n *such that we have*:

1) C *is the direct product of three cyclic groups of order* 2^n;
2) *The isomorphism class of* G *is determined by* n *and by whether* G *does or does not split over* C.

COROLLARY 1. *If S is a Sylow 2-subgroup of G then C has a basis u,v,w such that S is generated by C and elements s and t where*

$$t^2 = 1,\ u^t = w^{-1},\quad v^t = v^{-1},\quad w^t = u^{-1},$$

$$s^t = s^{-1},\quad u^s = v,\ v^s = w,\ w^s = uv^{-1}w$$

and

$$s^4 = 1 \text{ or } s^4 = uw$$

according as G splits or does not split over C.

It is interesting to note that if n = 1 then S is isomorphic with the Sylow 2-subgroup of A_8 or M_{12}, as G splits over C or does not. Moreover, as has been pointed out to me, if n = 2 and G does not split over C then S is isomorphic with the Sylow 2-subgroup of the Higman-Sims group.

COROLLARY 2. *There are involutions in G not in C.*

COROLLARY 3. *The sectional 2-rank of G is five.*

COROLLARY 4. *The 2-rank of G is three or four according as G does not split over C or does split over C.*

We now turn our attention to the fusion simple group described at the beginning and, with all that notation, state some more results.

THEOREM 2. *If the Sylow 2-subgroup of A has order at most two then A is of odd order and S is abelian or a Suzuki 2-group of order* 2^6 *or* 2^9.

THEOREM 3. *If the Sylow 2-subgroup of A has order four then A is isomorphic with* A_4 *and A has no non-identity fixed points in E.*

Whether this last case occurs has not yet been determined. However, we do know more. For example, there is a 3-element x in the normalizer of S and E with x not in the centralizer of E but with x^3 in the centralizer of S. Hence, $F = C_S(x)$ is cyclic or generalized quaternion. The case of F of order two has been analyzed in more detail.

Finally, using the above results, we have

THEOREM 4. *If every 2-local subgroup of G is solvable and P is a Sylow p-subgroup, for an odd prime p, of the automizer of a 2-subgroup of G then the following statements hold:*

1) *p is 3 or 7;*

2) *If p = 7 then S is abelian or a Suzuki 2-group;*

3) *If p = 3 then P is elementary abelian of rank at most three.*

University of Chicago

THE EXTENDED ZJ-THEOREM

J. L. Alperin, Richard Brauer, and Daniel Gorenstein

The extended form of Glauberman's ZJ-theorem stated in Theorem 2.7.2 of "Finite groups with quasi-dihedral and wreathed Sylow 2-subgroups", Trans. Amer. Math. Soc. 151 (1970), 1-162, has turned out to be an important tool in other classification problems of simple groups in terms of the structure of their Sylow 2-subgroups. However, it is not possible to deduce this result from the definition of p-*stability with respect to a* p-*group* as this notion is defined in Section 2.6 of that paper. The problem is that the required argument involves an implicit use of the Frattini argument that can not be justified in terms of the given definition of relative p-stability. We wish to thank David Goldschmidt for pointing out this error to us.

Because of the importance of this result, we wish to give here a modified definition of p-stability with respect to a p-group which is strong enough to allow one to deduce the extended ZJ-theorem. The modification consists, in effect, of incorporating the implicit use of the Frattini argument into the very definition. Thus the new definition involves three conditions, whereas the old one involved only two.

DEFINITION. Let H be a group in which $O_p(H) \neq 1$, p an odd prime, and let R be a p-subgroup of H such that

(a) $R \cap O_{p',p}(H)$ is a Sylow p-subgroup of $O_{p',p}(H)$;

(b) Either R is normal in a Sylow p-subgroup of H or RK/K contains $O_p(H/K)$ for every normal subgroup K of H;

(c) $H = KN_H(K \cap R)$ for every normal subgroup K of H.

Under these conditions we say that H is p-_stable with respect to_ R provided for any nontrivial subgroup P of R such that $O_{p'}(H)P$ is normal in H, we have

$$AC_H(P)/C_H(P) \subseteq O_p(N_G(P)/C_G(P)$$

for every p-subgroup A of R for which [P,A,] = 1.

Condition (c) is the added condition.

On the basis of this definition one proves first

PROPOSITION. _If_ B _is a nontrivial normal_ p-_subgroup of the group_ G _which is_ p-_stable with respect to the_ p-_subgroup_ P _of_ G, p _an odd prime, then_ $B \cap Z(J(P))$ _is normal in_ G.

As a corollary of this result, one obtains the desired extended form of Glauberman's ZJ-theorem.

THEOREM. _If_ G _is a group with_ $O_p(G) \neq 1$, p _an odd prime, which is_ p-_constrained and_ p-_stable with respect to the_ p-_subgroup_ P _of_ G, _then we have_

$$G = O_{p'}(G)N_G(Z(J(P))).$$

University of Chicago

Harvard University

Rutgers University

FINITE GROUPS GENERATED BY ODD TRANSPOSITIONS

Michael Aschbacher

THEOREM 1. *Let* G *be a finite group generated by a conjugacy class* D *of involutions such that for any pair* u, v, *of noncommuting elements in* D, uv *has odd order. Assume* G *has no nontrivial solvable normal subgroup and* $G' = G''$. *Then* G *is one of the following*:

(1) $Sp_n(q)$, $U_n(q)$, *or* $O_n^{\varepsilon}(q)$, q *even*.

(2) ${}^{\pi}O_n^{\varepsilon}(q)$, $q = 3$ *or* 5.

(3) $Sz(q)$.

(4) S_n.

(5) *one of the* 3 *Fischer groups*.

(6) $L_2(q)$ *wreath* S_n, q *even*.

Sp, U, O, Sz, S denote symplectic, unitary, orthogonal, Suzuki, symmetric groups, respectively.

Let ω be a set of integers and G a finite group. As defined by B. Fischer, a subset D of involutions of G is a *set of* ω*-transpositions* if $G = \langle D \rangle$, $D^G = D$, and $|uv| \in \{1,2\}\ \ \omega$ for all $u,v \in D$. D is a *set of odd transpositions* if ω is the set of all odd integers.

In [3] Fischer classifies groups generated by 3-transpositions. Theorem 1 classifies groups generated by odd transpositions.

There are equivalent formulations of the hypothesis of Theorem 1:

LEMMA 2. *Let* G *be a finite group and* D *a set of involutions with* $G = \ D$ *and* $D^G = D$. *Let* $P(u,v)$ *denote the property that* $C_D(uv) = C_D(u) \ \ C_D(v)$. *Then the following are equivalent*:

(1) D *is a set of odd transpositions of* G.

(2) *for all* $u \neq v$ *in* D, $P(u,v)$.

(3) *for all* u,v *in* D *with* $1 \neq uv = vu$, $P(u,v)$.

Therefore one can determine whether D is a set of odd transpositions of a group G by examing $C_G(uv)$ for involutions uv with u,v in D.

Let $G = Sp_n(q)$, q even, acting on a vector space V over GF(q) with symmetric bilinear form f. Let D be the set of transvections. A transvection u is determined by its center $u^* \varepsilon V$. $\langle u,v\rangle$ is determined by the 2-dimensional subspace U spanned by u^* and v^*. f restricted to U is degenerate exactly when u and v commute. If f restricted to U is nondegenerate then we can consider $\langle u,v\rangle$ to be a subgroup of the group of automorphisms of U: $SL_2(q)$.

This is a motivation for Theorem 1. Further it suggests a generalization for primes p > 2. The analogue for Theorem 1 and $p \geq 5$ appears in [1]. A special case for p = 3 appears in [2]. Fischer's theorem is the case q = 2.

Let D be a set of odd transpositions of G and u,v ε D. Then either $[u,v] = 1$ or u is conjugate to v in $\langle u,v\rangle$. This motivates the following series of definitions:

Let G be a finite group and D a collection of subgroups of G with $G = \langle D\rangle$ and $D^G = D$. D is *locally conjugate* in G if for A,B D either $[A,B] = 1$ or A is conjugate to B in $\langle A,B\rangle$. G acts on D by conjugation. Denote this action by G^D. A *set of imprimitivity* α for G^D is a subset of D with $\phi \neq \alpha \neq D$ and $\phi = \alpha \cap \alpha^g$ for

$g \in G - N_G(\alpha)$. Define

$$D_\alpha = \{\beta \in \alpha^G: [\alpha,\beta] = 1 \text{ and } \alpha \neq \beta\}.$$

Define a graph $\mathcal{D}(\alpha^G)$ with vertex set α^G and edges (α^g,α^h) where $\alpha^g \in D_\alpha h$. Let $D^*_\alpha = \{B \in \beta \in D_\alpha\}$.

LEMMA 3. *Let D be locally conjugate in G and α a set of imprimitivity for G^D. Let $A \in \alpha$, $B \in \alpha^g \neq \alpha$, and $[A,B] = 1$. Then $[\alpha,\alpha^g] = 1$.*

So $\mathcal{D}(D)$ and $\mathcal{D}(\alpha^G)$ are essentially the same graph.

THEOREM 4. *Let D be locally conjugate in G with G^D transitive and $\mathcal{D}(D)$ connected. Let $A \in D$. Then there is a unique maximal set of imprimitivity α of G^D containing A. Further $\langle D_\alpha \rangle$ is transitive on D^*_α and $\langle \alpha, D_\alpha \rangle$ is a maximal D-subgroup of G.*

So the proof of Theorem 1 comes in two parts. First determine groups G with $\mathcal{D}(D)$ disconnected. $\mathcal{D}(D)$ is disconnected exactly when $G' \neq G''$ or $G/O_\infty(G) \cong L_2(q)$, $Sz(q)$, or $U_3(q)$.

The second step is to consider G with $\mathcal{D}(D)$ connected and characterize G by $H = \langle D_\alpha \rangle$. By induction H is a known group. Further $\mathcal{D}(\alpha^G)$ is a graph extension of $\mathcal{D}(D_\alpha)$ with $H \leq G_\alpha$ transitive on D_α, the set of vertices adjacent to α.

Finally there is the following application of Theorem 1, motivated by the discussion above:

THEOREM 5. *Let G^Ω be a rank 3 permutation group, let $\alpha \in \Omega$, $H = G_\alpha$, and Δ a nontrivial orbit of H on Ω. Let D be the set of involutions conjugate in G to an element of H_Δ. Assume D is nonempty and any two points of Ω determine a unique line in the geometry associated with G and Δ. Then D is a class of odd transpositions of $\langle D \rangle$.*

The assumption on lines in Theorem 5 is weak and probably unnecessary.

REFERENCES

[1] M. Aschbacher, A characterization of the unitary and symplectic groups over finite fields of characteristic at least 5. (to appear)

[2] ____________, and M. Hall Jr., Groups generated by a class of elements of order 3. (to appear)

[3] B. Fischer, Finite groups generated by 3-transpositions. (to appear)

California Institute of Technology

GROUPS GENERATED BY A CLASS OF ELEMENTS OF ORDER 3.*

Michael Aschbacher and Marshall Hall, Jr.

1. INTRODUCTION.

The Conway group which is the group of automorphisms of the 24-dimensional Leech lattice is generated by a class of elements of order 3 with the property that any two of them either commute or generate $SL_2(3)$, $SL_2(5)$ or the alternating groups A_4, A_5 which are isomorphic to $SL_2(3)$ and $SL_2(5)$ modulo a center of order 2. Such a class of order 3 is also a special case of John Thompson's quadratic pairs for the prime 3.

In this paper our attention is restricted to elements of order 3 in which any two either commute, generate A_4 or $SL_2(3)$. It is possible to describe these groups completely.

Specifically the following theorem is proved:

THEOREM. *Let G be a finite group generated by a conjugacy class D of subgroups of order three, such that for any pair of noncommuting subgroups A and B in D, the group generated by A and B is isomorphic to $SL_2(3)$ or A_4. Assume G contains no nontrivial solvable normal subgroup. Then G is isomorphic to $Sp_n(3)$, $U_n(3)$, or $PGU_n(2)$, and D is a uniquely determined class of subgroups.*

2. LOCAL CONJUGACY

Let G be a finite group and D a collection of subgroups of G. Then D is said to be *locally conjugate*, if for any two subgroups A and B of D, either (i) A and B commute or (ii) A and B are conjugate in <A,B>. The groups of order three being investigated here have this property. A subgroup H of G is a D-subgroup if $H = \langle H \cap D \rangle$.

We consider the representation of G by conjugation on D. A <u>set of imprimitivity</u> Δ for this representation G^D is a subset of D such that $\phi \neq \Delta \neq D$ and for $g \in G$ either $\Delta = \Delta^g$ or $\Delta \cap \Delta^g = \phi$. A useful property is the fact that if $A \in \Delta$ and $B \in \Delta^g$ and if $[A,B] = 1$ then $[\Delta,\Delta^g] = 1$.

We define a graph $\mathcal{D}(\Delta^G)$ with point set Δ^G and edges (Δ^g,Δ^h) if $[\Delta^g,\Delta^h] = 1$. If D is a conjugate class and if the graph is connected then there is a unique maximal set of imprimitivity containing a given group A of D.

Let us introduce the following notation:

$$D_\alpha = \{\beta \in \alpha^G : [\alpha,\beta] = 1, \beta \neq \alpha\}$$
$$D^*_\alpha = \{B : B \in \beta \in D_\alpha\}$$
$$\alpha = \{\alpha\} \cup D_\alpha$$
$$A = \alpha^G - \alpha .$$

If the graph $\mathcal{D}(\alpha^G)$ is connected then a remarkable fact is that $\langle D_\alpha \rangle$ acts transitively on D^* by conjugation. If some further conditions hold, which are indeed by our class D of groups of order 3, then the following lemma holds.

LEMMA. G <u>is a rank</u> 3 <u>group acting by conjugation on</u> α^G, <u>the orbits of a stabilizer</u> G <u>being</u> $\{\alpha\}$, D_α, A_α.

Here it is to be understood that if D_α is vacuous then G is doubly transitive.

3. PARTICULAR GROUPS GENERATED BY A CLASS OF GROUPS OF ORDER 3

If $A = \langle a \rangle$ and $B = \langle b \rangle$ are subgroups of order 3 which do not commute but $\langle A,B \rangle = SL_2(3)$ or A_4 then choosing the generators appropriately we have

$$a^3 = b^3 = 1,\ aba = bab.$$

We write $a \sim b$ for these relations. There are four D subgroups, namely $\langle a\rangle$, $\langle b\rangle$, $\langle a^{-1}ba\rangle$, $\langle b^{-1}ab\rangle$, and any two of these generate $SL_2(3)$. Note that we do not have $a \sim b^{-1}$, though we do have $a^{-1} \sim b^{-1}$. In $SL_2(3)$, $(a^{-1}b)^2$ is an involution in the center while in A_4 we have $(a^{-1}b)^2 = 1$, and A_4 can be defined by

$$a^3 = b^3 = 1,\ (a^{-1}b)^2 = 1.$$

The three generator groups from the class D, apart from the case in which all three commute, contain two, $A = \langle a\rangle$ and $B = \langle b\rangle$ in which $a \sim b$. Then if $C = \langle c\rangle$ is a further D group one of the following must hold:

3.1 $a \sim b$ in all cases. $G = \langle a,b,c\rangle$.

3.2 $ca = ac$, $cb = bc$. Here $G = \langle a,b\rangle \times \langle c\rangle$.

3.3 $ca = ac$, $c \sim b$. Here $|G| = 648$
and putting $h = a^{-1}c$, G has a normal subgroup $H = \langle h, b^{-1}hb\rangle$ of exponent 3 and order 27 and $G/H \simeq \langle a,b\rangle$. Here G has 12 subgroups in D.

3.4 $c \sim a$, $c \sim b$, $c \sim a^{-1}ba$, $c \sim b^{-1}ab$.
Here $|G| = 768 = 2^8 3$. G has 16 subgroups in D. G has a center Z which is elementary abelian of order 16 and $Z = \langle (a^{-1}c)^2, (a^{-1}c)^2, (b^{-1}c)^2, (a^{-1}b^{-1}ac)^2\rangle$. Here G/Z has a normal subgroup K/Z of index 3 which is elementary abelian and generated by conjugates of $a^{-1}b$.

3.5 $c \sim a$, $c \sim b$, $c \sim a^{-1}ba$, $c^{-1} \sim b^{-1}ab$.
These relations make G collapse so that $G = 1$.

3.6 $c \sim a, c \sim b, c^{-1} \sim a^{-1}ba$, $c^{-1} \sim b^{-1}ab$.
Here $|G| = 6048$ and G is isomorphic to the simple group $U_3(3)$. G contains 28 subgroups in D, which lie four at a time in 63 $SL_2(3)$'s.

A critical theorem is the following:

THEOREM. *If* G *is generated by* D *groups no two of which commute then either* $G = U_3(3)$ *or* $|G| = 2^n \cdot 3$ *and* G *has a normal series* $G > H > K \geq 1$ *where* G/H *is of order* 3 *and* H/K *and* K *are elementary abelian* 2-*groups,* K *being in the center of* G.

In proving this if we assume that any three distinct D groups, not in an $SL_2(3)$, say $\langle a \rangle$, $\langle b \rangle$, and $\langle c \rangle$ are such that $\langle a,b,c \rangle$ is of type 3.4 above then $(a^{-1}b)^2$ commutes with c and so is in the center of G. This easily leads to the solvable group of order $2^n \cdot 3$ described above. But if $\langle a,b,c \rangle$ is $U_3(3)$ it is necessary and fairly difficult to show that a further D group $\langle t \rangle$ necessarily commutes with at least one of the 28 D groups in $U_3(3)$. Here t together with each of the 63 $SL_2(3)$'s of $U_3(3)$ must be of type 3.4 or type 3.6 above and type 3.5 must be avoided. An analysis of these cases shows that there are a number of type 3.4 and as t commutes with the central involution of the $SL_2(3)$ in each case, this leads to the collapse to the identity of the group $\langle a,b,c,t \rangle$. If t commutes with exactly one of the 28 D groups in $U_3(3)$, then $\langle a,b,c,t \rangle = G$ has a kernel K which is a 3-group and $G/K \simeq U_3(3)$. If t commutes with more than one of the 28 D groups (but not all) then t commutes with the 4 groups in an $SL_2(3)$ and $\langle a,b,c,t \rangle$ is $U_4(3)$ together with its Schur multiplier $Z_3 \times Z_{12}$.

4. THE CLASSICAL GROUPS OF THE MAIN THEOREM

Let $A = \langle a^* \rangle$ and $B = \langle b^* \rangle$ be two non-commuting D groups and for $X \in D$ defining $D_X = \{C: C \in D, [X,C] = 1, C \neq X\}$ a further group L is defined by $L = \langle D_A \cap D_B \rangle$.

Case I. $Sp_{2m}(3)$. Let V be a 2m-dimensional symplectic space over GF(3) with non-degenerate skew symmetric bilinear form (,). For $u \in V^{\#}$ the transvection u^* determined by u is the map

$$u^* : \langle x \rangle \to \langle x + (x,u)u \rangle$$

considered as a projective transformation of V. Let $D = \{\langle u^* \rangle : u \in V^{\#}\}$ and $G = \langle D \rangle$.

Here G is the 2m-dimensional projective symplectic group $Sp_{2m}(3)$ over GF(3).

In this case for $A = \langle a^* \rangle$ and $B = \langle b^* \rangle$ in D if $[A,B] \neq 1$, then

(1) $\langle A,B \rangle = SL_2(3)$ or A_4

(2) $L/Z(L) = Sp_{2m-2}(3)$ for $m > 1$.

Case II. $U_n(3)$. Let V be an n-dimensional unitary space over GF(9) with non-degenerate semibilinear form (,). For a nonzero singular vector u of V let u^* be the transvection determined by u considered as a projective transformation of V. Let $D = \{u^* : u \in V, (u,u) = 0\}$ and $G = \langle D \rangle$.

Here G is the n-dimensional projective special unitary group $U_n(3)$. In this case if $A, B \in D$, and $[A,B] \neq 1$, then

(1) $\langle A,B \rangle = SL_2(3)$ or A_4

(2) $L/Z(L) = U_{n-2}(3)$

(3) G contains a unique class of D-subgroups K^a with $K/Z(K) = U_{n-1}(3)$.

Case III. $PGU_n(2)$. Let V be an n + 1 dimensional unitary space over GF(4) with nondegenerate semibilinear form (,). For nonsingular $u \in V$ let u^* be the group generated by the projective transformation

$$\langle x \rangle \to \langle x + \alpha(x,u)u \rangle$$

where α is a generator of the multiplicative group of GF(4). Put $D = \{u^* : (u,u) = 1\}$ and $G = \langle D \rangle$.

G is the projective general unitary group $PGU_{n+1}(2)$. $[PGU_n(2) : U_n(2)] = 3$ if n is divisible by 3, while the two groups are equal otherwise.

In this case, with $A,B \in D$ and $[A,B] \neq 1$, putting $H = \langle D_A \rangle$, $L = \langle D_A \cap D_B \rangle$, then

(1) $\langle A,B \rangle = SL_2(3)$ or A_4

(2) $H = GU_n(2)$

(3) $L/O_2(L) = PGU_{n-2}(2)$ for $n \geq 4$. $O_2(L)$ has order 2^{2n-3} and is in the center of L

(4) H contains a unique class of D-subgroups satisfying the properties of L in (3).

5. PROOF OF THE MAIN THEOREM

In the proof we consider a group G which is a counterexample of minimal order to the main theorem. This argument is essentially the same as an inductive argument.

First it can be shown that G' is simple and of index at most 3. Then we can reduce to the case in which the graph is connected and the maximal set of imprimitivity α consists of a single group A.

From the theorem in Section 3 the cases in which D_A is vacuous are known. Hence we put $H = \langle D_A \rangle$ and define as the maximal normal solvable subgroup of H. By induction H is known and using its properties and D. Higman's relations on rank 3 groups, with a considerable amount of work it can be shown that G must be one of the groups of the theorem.

Much of the argument on conjugacy derives from reference [1], while the material in Section 3 comes from reference [3].

* This research was supported in part by ONR contract N00014-67-A0094-0010.

REFERENCES

[1] M. Aschbacher, On finite groups generated by odd transpositions, (to appear).

[2] P. Dembowski and A. Wagner, Some characterizations of finite projective space. Arch. Math. 11 (1960), 465-469.

[3] M. Hall, Jr., Combinatorial designs and groups, Proceedings of the International Congress of Mathematicians, Nice, France, 1970.

[4] D. Higman, Finite permutation groups of rank 3, Math. Zeitschr. 86 (1964), 145-156.

[5] R. Steinberg, Automorphisms of finite linear groups, Canadian J. Math. 13 (1960), 606-615.

California Institute of Technology

SOLVABLE GROUPS, AUTOMORPHISM GROUPS, AND REPRESENTATION THEORY*

T. R. Berger

In [3] the following theorem is proved.

THEOREM 1. *If a solvable group G admits a nilpotent $Z_p \wr Z_p$ free (for all primes p) fixed point free group of operators A (i.e. $C_G(A) = 1$) where $(|A|,|G|) = 1$ then the Fitting height of G is bounded above by the composition length of A.*

If A is abelian then it is certainly $Z_p \wr Z_p$ free. In [13] Shult showed Theorem 1 held for abelian A if we excluded the possibility that $d = r^t + 1$ for divisors $d \mid \exp A$ and primes r with $r^m \mid |G|$. There he mentioned that in the prime exception case the bound might double. Of course Theorem 1 shows that this is not so and Shult's original theorem holds without prime exceptions.

In [2] the fact that a class 2 odd p-group is regular was used. In [10] a distinction appears when the operator group is regular or not. A regular p-group is clearly $Z_p \wr Z_p$ free. Thus Theorem 1 settles all cases when A has regular Sylow subgroups.

This theorem in turn depends upon the following representation theorem proven in [1].

THEOREM 2. *Assume AG is a solvable group with normal subgroup G and nilpotent $Z_p \wr Z_p$ free (for all prime p) complement A where $(|A|,|G|) = 1$. Let k be a field of characteristic prime to $|A|$ and V a faithful irreducible k[AG]-module. Suppose $R \leq G$ is a normal*

* This research was partially supported by NSF grant GP 29224X.

r-subgroup of AG *such that*

(1) R *contains a unique maximal* AG-*invariant subgroup* R^*, *and*

(2) *there is an* AG-*cheif factor* $L/C_L(R/R^*)$ *of* G *so that* $L = C_L(X/Y)$ *for all* AG-*cheif factors of* R^*.

If $A = A^0 > \dots > A^n = 1$ *is a central compositions series for* A *and* $C_V(A^m) = (0)$ *but* $C_V(A^{m+1}) \neq (0)$ *then*

(a) $C_{R/R^*}(A^m) = C_{R/R^*}(A^{m+1})$ *and*

(b) $C_{G/F(G)}(A^m) = C_{G/F(G)}(A^{m+1})$.

Earlier theorems like Theorem were proved in [2, 10, 13] for A variously abelian, odd p class ≤ 2, and odd p-groups. Also certain numerical equations on numbers dividing $|A|$ and $|G|$ had to be satisfied. These fixed point free theorems were proved by representation theorems resembling Theorem 2. Thus the quest is for general representation theorems like Theorem 2. In a proof of Theorem 2 the following two results are crucial.

Assume AG is a solvable group with normal subgroup G and nilpotent $Z_p \wr Z_p$ free (for all primes p) complement A where $(|A|,|G|)=1$. Let k be a field of characteristic prime to $|A|$. Finally suppose V is a faithful irreducible k[AG]-module.

THEOREM 3. *If* $k = GF(r)$ *for some prime* r *then* A *permutes the elements of* V *with at least one regular orbit occurring*.

THEOREM 4. *If* k *is algebraically closed and* $R \leq G$ *is a normal extraspecial subgroup of* AG *such that* $Z(R) \leq Z(AG)$, $C_A(R) = 1$, *and* $R/Z(R)$ *is irreducible as an* AG-*module then* $V|_A$ *contains a regular* A *direct summand or*

(1) $G = C_G(R/Z(R))$ *and*

(2) A *is cyclic or a p-group for some prime* p.

Proofs of these two results appear in [9].

Let us discuss for a moment an apparently unrelated topic. Suppose C is a cyclic p-group for some prime p. Assume R is a special r-group for a prime $r \neq p$. Suppose CR is a group with R as a normal subgroup, $R' \leq Z(CR)$ and R/R' is a faithful irreducible C-module. Let k be a field of characteristic p and let V be a faithful irreducible CR-module. This is precisely the situation occuring in Theorem B of Hall and Higman [12].

Suppose $C = \langle w \rangle$. Hall and Higman were concerned with conditions guaranteeing that w would have minimal polynomial $x^{|C|} - 1$ on V. Equivalently they wanted to know under what conditions $V|_C$ had a regular C-direct summand. Two possibilities arise.

First is the case where R is elementary abelian. Here R is a vector space over GF(r) and C acts faithfully and irreducibly upon R. In fact, as a C-module C permutes the non-identity elements of R in all regular orbits. This fact is used to completey Theorem B in this case [11 pp. 73, 362]. That is, the fact that C acts semi-regularly upon $R^{\#}$ is of some importance. Actually it may be possible to obtain some information from just one regular orbit of C on R. With C in place of A, G = 1, and R = V we see then this case resembles Theorem 3.

Second, the case where R is non-abelian, i.e. R is extraspecial. For Theorem B we need to know when $V|_C$ has a regular direct summand. With A = C, R = G we see that this resembles Theorem 4.

The point of all this is to say that the following two questions are of some importance in widely differing settings.

Suppose AG is a solvable group with normal subgroup G and nilpotent complement A where $(|A|,|G|) = 1$. Assume k is a field and V is a faithful irreducible k[AG]-module. The two situations outlined above are special cases relating to the following two questions:

(1) *If* $k = GF(r)$, *when does the permutation representation of* A *upon* $V^{\#}$ *contain a regular*?

(2) *Let* S G *be a normal extraspecial* s-*group of* AG *with* $Z(S) \leq Z(AG)$, $C_A(S) = 1$, *and* $S/Z(S)$ *irreducible as an* AG-*module*. *When does* $V|_A$ *contain a regular* A-*direct summand*?

The condition that $C_A(S) = 1$ is not unduly strong. In [1] appear results showing ways to remove this condition.

A systematic study of these two questions has been started in [4,5,6,7,8,9]. Let us return again to the Theorem B case of Hall and Higman. If R is extraspecial and k is algebraically closed then $V|_C \simeq mW \dotplus J$ where W is the regular C-module and J is the indecomposable of dimension $|C| - 1$. Thus $W \simeq k[C]^+$ and $J \simeq k[C]^+/Kk[C]$ where $K = \sum_{x \in C} x$. When k is not of characteristic p, $V|_C \simeq mW \dotplus J$ where again W and J are isomorphic to the same sections of $k[C]$. That is, the modular and ordinary theory for (2) in this case is the "same". It would appear that there might be some systematic way of attacking (1) and (2) independent of field characteristic. This is started in [7].

It is not obvious but there is a great deal of similarity between (1) and (2). Question (1) may be called a problem in additive module structure. Then question (2), viewed in the same way, becomes a problem in multiplicative module structure.

Let us look more closely at (1). We may find subgroups $A_o \leq A$, $G_o \leq G$ so that A_oG_o is a group and there is a primitive

A_oG_o-module U such that $U|^{AG} \simeq V$. To give a complete answer to (1) then it is only necessary to answer the question for primitive configurations (A_oG_o, U) and then determine completely what the process of induction does to things. So the question breaks into two parts: (a) analysis of minimal structures, and (b) application of induction. The case (b) involbes an analysis of additive module structure.

The question (2) breaks up similarly. The analysis of minimal cases considers certain "primitive" (not quite primitive) actions of AG upon S/Z(S). The application of induction in the multiplicative module structure of AG. This multiplicative structure case is defined in [8]. Actually the second part (b) of both questions (1) and (2) are so similar that they can be treated together [8]. We shall refer to this as the inductive step.

The process of answering (1) and (2) involves starting with noncomplicated cases and slowly building up to more complex situations. In this process there is a certain amount of "feedback". That is, answers to (2) for certain cases are needed to give more complex answers to (1) and vice versa.

The question (1) seems to precede in this process. This can be stated as follows. Suppose we can answer (1) under some hypothesis P on AG and V. With the same hypothesis P and a few additionalconditions. For example, in going from Theorem 3 to Theorem 4 conditions (1) and (2) have been added onto the conclusion. The condition that k be algebraically closed is just incidental.

The necessity for additional hypothesis or conditions in (2) occurs both because of the inductive step and also because of minimal cases.

REFERENCES

[1] T. R. Berger, Automorphisms of solvable, (to appear).

[2] ___________, Class two p-groups as fixed point free automorphism groups, Ill. J. Math. 14 (1970), 121-149.

[3] ___________, Nilpotent fixed point free automorphism groups of solvable groups, (to appear).

[4] ___________, Hall- Higman type theorems I, (to appear).

[5] ___________, Hall-Higman type theorems II, (to appear).

[6] ___________, Hall-Higman type theorems III, (to appear).

[7] ___________, Hall-Higman type theorems IV, (to appear).

[8] ___________, Hall-Higman type theorems V, (to appear).

[9] ___________, Hall-Higman type theorems VI, (to appear).

[10] ___________, Odd p-groups as fixed point free automorph-groups, Ill J. Math. 15, (1971), 28-36.

[11] D. Gorenstein, Finite Groups, Harper and Row, New York, 1968.

[12] P. Hall and G. Higman, On the p-length of p-soluble groups and reduction theorems for Burnside's problem, Proc. London Math. Soc. (3) 6 (1956), 1-42.

[13] E. E. Shult, On groups admitting fixed point free abelian operator groups, Ill. J. Math. 9 (1965) 701-720.

GROUPS WHOSE SYLOW 2-GROUPS HAVE CYCLIC COMMUTATOR GROUPS

Paul Chabot

The 2-groups D_{2^n} (dihedral), S_{2^n} (semi-dihedral), $Z_{2^n} \wr Z_2$ (wreath product) do appear in known simple groups as Sylow 2-groups. Moreover, all groups having such a Sylow 2-group have been classified. Each of these 2-groups has a cyclic commutator group. One can thus ask if these are the only 2-groups possessing cyclic commutator groups which can appear as Sylow 2-groups of simple groups. In answer to this we have the following partial answer.

THEOREM. *Let* $P \in Syl_2(G)$ *and suppose the commutator group* P' *is cyclic. Then, one of the following holds*:

i) $O_{2',2}(G) > O_{2'}(G)$,

ii) $O^2(G) < G$

iii) $P \simeq A \times B$, *where* A *is elementary and* B *is one of* D_{2^n}, S_{2^n}, *or* $Z_{2^n} \wr Z_2$.

The proof is entirely fusion analysis. It draws heavily upon Glauberman's Z*-theorem [2], Goldschmidt's $\mho^1$-theorem [3], the focal subgroup theorem, and particularly the theorem of Alperin concerning fusion [1]. Alperin's theorem is very often applied as follows:

If $\gamma \in Z(P)$ *is fused to* $\beta \notin Z(P)$, *then there exists a* $\gamma' \in Z(P)$ *fused to* γ, *a* $\beta' \notin Z(P)$, *a Sylow intersection* $P \cap Q = C_P(\beta')$, *and a 2-element* $x \in N(P \cap Q)$ *such that* $x: \beta' \to \gamma'$.

Let τ be the involution of $P' = \langle\sigma\rangle$. In most cases it is proved or known that τ is isolated in $P_0 = C_P(\sigma)$. This fact is then employed to determine that $Z(P)$ is elementary abelian as follows:

Let $\gamma \in \Omega_1(\mho^1(Z(P)))$ and suppose γ is fused to $\beta \notin Z(P)$. By using Alperin's theorem we obtain γ', β' and $x \in N(C_P(\beta'))$. Since β' must then have a root in P we get that $\beta' \in P_0$. But then we observe that $[\beta', x] = \beta'\gamma'$ must be fused to τ (since it is an involutive commutator of the 2-group $C_P(\beta') \langle x \rangle$). Since is isolated in P_0, we have a contradiction: This means that $\Omega_1(\mho^1(Z(P)))$ is weakly closed in $Z(P)$ and hence trivial by Goldschmidt's theorem.

Once this point is reached, the structure of P is fairly restricted and the focal subgroup theorem is applied.

REFERENCES

[1] J. L. Alperin, Sylow intersections and fusion, J. Algebra 6 (1967), 222-241.

[2] G. Glauberman, Central elements in core-free groups, J. Algebra 4 (1966), 403-420.

[3] D. Goldschmidt, On the 2-exponent of a finite group, Ph. D Thesis, University of Chicago, 1969.

Los Angeles State College

A CONSTRUCTION FOR THE SMALLEST FISCHER GROUP F_{22}

J. H. Conway

The Fischer groups F_{22}, F_{23}, F_{24} are closely related to the corresponding Mathieu groups, and it is natural to attempt constructions in terms of these. In this paper we show that certain difficulties with F_{23} and F_{24} do not arise for the smallest group F_{22}.

1. THE GROUP M_{24}.

M_{24} is a quintuply transitive group on a set Ω of 24 letters. Defining the sum of two subsets of Ω as their symmetric difference, we obtain a 24-dimensional vector space over GF(2), in which M_{24} leaves invariant a 12-dimensional subspace **C** (the Golay 24-code) of **C**-sets, namely ϕ and Ω , together with 759 8-element sets called octads and their complements, and 2576 12-element sets called dodecads.

We obtain the subgroup M_{22} by fixing two points 0 and ∞ of Ω. We shall use the facts that there are just 77 hexads which are completed to octads by $\{0,\infty\}$, and that any one hexad is disjoint from just 16 others. Two disjoint hexads define a partition of Ω as $2 + 6 + 6 + 10$ in such a way that the union of any two parts is a **C**-set, and the stabiliser of this partition in M_{24} is a group S_6 whose action we describe in some detail.

In the group S_6 which permutes 6 letters i,j,k,l,m,n there are just 12 subgroups of index 6, namely the 6 subgroups S_5 fixing one letter each, and 6 further subgroups which permute {i,j,k,l,m,n} in the way that $PGL_2(5)$ permutes the symbols $\{\infty,0,1,2,3,4\}$.

(Namely by linear fractional transformations $x \to (ax+b)/(cx+d)$.) The $PGL_2(5)$ in which i,j,k,l,m,n play the respective roles of ∞,0,1,2,3,4 will be called G(i|jklmn). It has in all 120 names of this kind, which can be obtained from

G(i|jklmn), G(j|ilknm), G(k|imljn), G(l|inmkj), G(m|ijnlk), G(n|ijkml)

by regarding the last 5 letters in each case as a 5-cycle which can be rotated or replaced by its powers. The S_6 has an outer automorphism which interchanges the 6 subgroups S_5 with the 6 subgroups $PGL_2(5)$.

Now partition the set Ω as 2 + 6 + 6 + 10 so that the union of any two parts is a C-set, and let the 2-element part be $\{0,\infty\}$ and the 6-element parts {i,j,k,l,m,n} and {u,v,w,x,y,z}. Then the subgroup S_6 of M_{24} which fixes the partition acts as follows:

i) Even permutations of S_6 fix 0 and ∞; odd permutations interchange them.

ii) The stabiliser of one of u,v,w,x,y,z is one of the 6 groups like G(i|jklmn).

iii) The stabiliser of any of the 10 remaining points fixes one of the 10 partitions 3 + 3 of {i,j,k,l,m,n}.

2. THE GROUP F_{24}

F_{24} has a conjugacy class of 306936 involutions called <u>transpositions</u>, any non-commuting pair of which have product of order 3. A maximal commuting set of these contains just 24, generating a group of order 2^{12} which is self-centralising in F_{24}. We call 24 such involutions the <u>basic transpositions</u>. The normaliser of the group they generate realises the permutations of M_{24} on them, the C-sets being sets of basic transpositions with trivial product. In fact this normaliser N is a non-split extension $2^{12}M_{24}$ whose orbits on transpositions are:

i) the 24 basic transpositions,

ii) a set of $2^5 \cdot 759$ transpositions each commuting with just 8 basic transpositions forming an octad (There are just 32 of these per octad.), and

iii) a set of $2^{10} \cdot (24 \cdot 23/2)$ transpositions each commuting with just 2 basic transpositions. (There are just 2^{10} of these per pair of basic transpositions.)

Since any 5 points of Ω belong to just one octad, any 4 points of Ω belong to just 5 octads, which are otherwise disjoint. It follows that the product of any 5 basic transpositions can be written as a product of 4 others in just 5 more ways.

3. THE GROUP F_{22}

To obtain F_{22} from F_{24} we factor the centraliser of two commuting basic transpositions 0 and ∞ by the subfroup of order 4 generated by 0 and ∞. It is then easy to see that in F_{22} we have the following transpositions:

i) the 22 remaining basic transpositions (or rather, their images in F_{22}).

ii) a set of $2^5 \cdot 77$ transpositions each commuting with just a hexad of the above 22.

iii) a set of 2^{10} transpositions commuting with no one of the basic 22.

Now fortunately, when we factor the (non-split) extension of 2^{12} by M_{22}, we obtain a split extension $2^{10}M_{22}$, for the subgroup fixing one of the transpositions in orbit iii) has index 2^{10}. It is this that makes it easy to construct the group F_{22}.

We define F_{22} in terms of the 22 basic transpositions i,j,k,.. (typically t), and a further transposition S taken from the 2^{10} orbit, and fices by a group M_{22}. The remaining transpositions of this orbit are written as conjugates of S by products of basic transpositions and the transpositions of the other orbit have names (ijk|lmn), (i|jklmn), and (ijklmn|uvwxyz). In each case, this transposition commutes with i,j,k,l,m,n.

In the table we specify the action of S and the typical basic transposition t on the entire set of 3510 basic transpositions, which suffices to define F_{22} since the basic transpositions together with S generate F_{22}. We now describe the orbits of M_{22} on the transpositions in more detail.

The typical member of the 2^{10} orbit of $2^{10}M_{22}$ can be written as the transform of S by a product of basic transpositions i,j,k,... We never need moe than three transpositions in the product because a product of 4 basic transpositions can be weitten as a product of another 4 including 0 or , which can be neglected. Again, if i,j,k are three of the 22 basic transpositions, we have 0ijk = lmn for just one triple l,m,n, since {∞,0,i,j,k} defines a unique octad. So the involution $S^{ijk} = S^{lmn}$ has just two names of this form, where {i,j,k,l,m,n} is a hexad.

Now if {i,j,k,l,m,n} is a hexad, there are just 32 transpositions which commute just with i,j,k,l,m,n. These are

i) 10 involutions (ijk|lmn), the transform of $S^{ijk} = S^{lmn}$ by S.

ii) 16 involutions (ijklmn|uvwxyz), say, which transform by S into involutions commuting with the members of the disjoint hexad {u,v,w,x,y,z}. These correspond one for one with the 16 such hexads disjoint from {i,j,k,l,m,n}.

TABLE 1.

NAME	M_{22} ORBIT LENGTH	$2^{10}M_{22}$ ORBIT LENGTH	ACTION OF S	ACTION OF A BASIC TRANSPOSITION t	
i	22	22	S^{i}	Fixed	
S	1		Fixed	S^{t}	
S^{i}	22	1024	i	S^{it}	
S^{ij}	231		Fixed	S^{ijt}	(Some of these will have shorter names)
$S^{ijk} = S^{lmn}$	770		$(ijk \mid lmn)$	$S^{ijkt} = S^{lmnt}$	
$(ijk \mid lmn)$	$10 \cdot 77$		$S^{ijk} = S^{lmn}$	Fixed <u>or</u> $(ijklmn \mid uvwxyz)$.	
$(i \mid jklmn)$	$6 \cdot 77$	$32 \cdot 77$	Fixed	Fixed <u>or</u> $(ijklmn \mid uvwxyz)$.	
$(ijklmn \mid uvwxyz)$	$16 \cdot 77$		$(uvwxyz \mid ijklmn)$	Fixed <u>or</u> $(i \mid jklmn)$ or $(ijk \mid lmn)$	

iii) 6 involutions (i|jklmn), say, corresponding one for one with the 6 subgroups G(i|jklmn) of the S_6 on {i,j,k,l,m,n}.

(The proof of these assertions is fairly easy. We know that a transposition commutes with just 693 other, so S cannot fix the 770 involutions $S^{ijk} = S^{lmn}$, and must therefore take them to a new M_{22} orbit (ijk|lmn), say, consisting of points commuting with <u>some</u> hexad of the basic 22. By symmetry considerations this hexad can only be {i,j,k,l,m,n}, and we have found 10 of the 32 transpositions commuting with these points. Next, it is easy to see that some involution commuting with i,j,k,l,m,n must be taken to one commuting with another hexad u,v,w,x,y,z. This cannot be one of the 60 hexads intersecting {i,j,k,l,m,n}, for then we should obtain 60 further transpositions. So we must obtain in this way just 16 new transpositions (ijklmn|uvwxyz), say. There remain just 32 - 10 - 16 = 6 further transpositions to be found, which must be associated <u>either</u> with the 6 points i,j,k,l,m,n <u>or</u> with the 6 subgroups like G(i|jklmn), since there are only two classes of subgroups of index 6 in the S_6 on {i,j,k,l,m,n}. The first possibility is easily disproved.)

We now know the action of S on most of the orbits, and can fill in the gaps by symmetry and the assertion that S fixes just 693 other transpositions. Moreover, we know the action of a typical basic transposition t on all points except those of the form (ijk|lmn) or (i|jklmn) of (ijklmn|uvwxyz) when t $\notin$ {i,j,k,l,m,n}. Symmetry considerations now force a unique action, which is best described by considering the various transforms of (ijklmn|uvwxyz). We shall need our knowledge of the stabiliser in M_{24} of the 2 + 6 + 6 + 10 partition whose parts are {0,∞},{i,j,k,l,m,n}, {u,v,w,x,y,z} and the 10 remaining points.

i) If t is one of i,j,k,l,m,n, then (ijklmn|uvwxyz) is fixed by t.

ii) If t is one of u,v,w,x,y,z, the transform is (i|jklmn), where G(i|jklmn) is the subgroup fixing t.

iii) Otherwise, the transform is (ijk|lmn), where {i,j,k}, {l,m,n} is the partition 3 + 3 of {i,j,k,l,m,n} which is fixed by the stabiliser of t.

These observations produce the last entry in the table, and also, since t is an involution, the entries above it, and so complete the construction of F_{22}. We have not attempted to give an independent existence proof for F_{22} using this construction, and so are still relying on Fischer's work. But such a proof would be quite easy: all that is needed is a description in these terms of which pairs of the transpositions commute, and verification that S preserves these pairs.

More interesting is the possibility of extending the construction to F_{23} and F_{24}, and the possible discovery of new subgroups of these groups using properties of the notation. I have already used the F_{22} construction to verify that the Tits simple group $({}^2F_4(2))'$ is contained in F_{22}. (It was asserted by Fischer and Rudvalis that the fixed point set of a suitable automorphism was of the same order as the Tits group, and it was not difficult to verify that indeed there was an automorphism of F_{22} which fixed certain elements satisfying the relations given by Tits for the generators of his group. Relations involving permutations of M_{22}, and the involutions S,t are easily checked using the table.)

In this sort of work it is convenient to have efficient ways of computing the actions of the various elements. This is easy for F_{22} if we use Robert Curtis's "MOG" (Miracle Octad Generator), which enable one quickly to find the octad determined by 5 points, together with the following rules for transforming the difficult elements.

We first study the relation between the two hexads {i,j,k,l,m, n} and {u,v,w,x,y,z} in more detail.

i) A <u>duad</u> (unordered pair) from one hexad corresponds to a <u>syntheme</u> (triple of disjoint duads) in the other. This correspondence is given by the identities

$\infty 0ij = klmn = uv?? = wx?? = yz?? = ????$ for $ij \leftrightarrow uv.wx.yz$

$\infty 0uv = wxyz = ij?? = kl?? = mn?? = ????$ for $uv \leftrightarrow ij.kl.mn$

in terms of the basic transpositions, where ? signs denote the unnamed transpositions.

ii) A <u>point</u> of one hexad corresponds to a <u>synthematic total</u> (set of 5 synthemes including all 15 duads) in the other. This correspondence is given by letting:

i map to the 5 synthemes obtained from ij,ik,il,im,in

u map to the 5 synthemes obtained from uv,uw,ux,uy,uz.

In terms of these concepts, the group $G(x_\infty | x_0 x_1 x_2 x_3 x_4)$ s more symmetrically defined as the stabiliser of the set of 5 synthemes:

$$x_\infty x_i . x_{i+1} x_{i-1} . x_{i+2} x_{i-2} \quad \text{(i running modulo 5)}$$

and we shall regard it as defined when we have specified 5 such synthemes.

EXERCISE. The group stabilising {ab.cd.ef, ac.be.df, ad.bf.ce, ae.bd.cf, af.bc.de} is G(a|bcefd). Verify this, and find in the same way its name G(f|ed???).

<u>To compute</u> $(ijklmn|uvwxyz)^t$ <u>when</u> t = u, find the 5 synthemes in {i,j,k,l,m,n} corresponding to uv,uw,ux,uy,uz.

<u>To compute</u> $(i|jklmn)^t$, $t \notin \{i,j,k,l,m,n\}$, let ij.kn.lm be one

of the 5 corresponding synthemes for G(i|jklmn). Locate the unique duad tv such that ∞0tv = ij?? = ij?? = kl?? = mn??. Similarly find w,x,y,z from the other synthemes. Then $(i|jklmn)^t$ = (ijklmn|uvwxyz), where u = t.

(NOTE. The equations determine v as the unique common point of the three triplets defined by the equivalent equations ∞0ijt = v??, ∞0klt = v??, ∞0mnt = v??,)

To compute $(ijklmn|uvwxyz)^t$ _when_ t _is one of the_ 10 _unnamed points_. Find the partitions 3 + 3 of numerator and denominator so that (with suitable names):

∞ijk = 0lmn = tuvw, 0ijk = ∞lmn = txyz.

Then $(ijklmn|uvwxyz)^t$ = (ijk|lmn).

To compute $(ijk|lmn)^t$, $t \notin \{i,j,k,l,m,n\}$.
Solve the same equations.

California Institute of Technology

GROUPS WITH A (B,N)-PAIR OF RANK 2

Paul Fong and Gary M. Seitz

A group G is said to have a (B,N)-pair if there exist subgroups B and N of G satisfying the following conditions:

(i) $G = \langle B,N\rangle$ and $H = B \cap N \trianglelefteq N$.

(ii) $W = N/H$ is generated by a set S of involutions.

(iii) $sBw \subseteq BwB \cup BswB$ for $s \in S$, $w \in W$.

(iv) $sBs \neq B$ for $s \in S$.

The rank of the (B,N)-pair is defined to be $|S|$. The standard examples of groups with (B,N)-pairs are the groups obtained by Chevalley as well as the variations of these groups obtained by Hertzig, Ree, Steinberg, Suzuki, and Tits. We will use the term Chevalley group to mean that subclass of the above class of groups that have trivial center and are generated by their root subgroups. The main result is a classification of finite groups with a (B,N)-pair of rank 2 satisfying an additional assumption.

THEOREM 1. *Let G be a finite group with a (B,N)-pair of rank 2 satisfying*

(*) *there exists a normal nilpotent subgroup U of B such that $B = HU$.*

Let $G_0 = U^G$, $B_0 = B \cap G_0$, $N_0 = N \cap G_0$, and $Z = \bigcap_{g \in G_0} B_0^{\,g}$.

Then one of the following holds:

a) $|W| = 4$ *and G_0/Z is solvable.*

b) $|W| = 4$ *and $G_0/Z = L_1 \times L_2$ where L_1 and L_2 are groups with*

(B,N)-pairs of rank 1 satisfying (*).

c) $|W| \neq 4$ *and* G_0/Z *is isomorphic to a Chevalley group of rank* 2, *that is*, $G_0/Z \cong PSL(3,q)$, $PSp(4,q)$, $PSU(4,q)$, $PSU(5,q)$, $G_2(q)$, ${}^3D_4(q)$, *or* ${}^2F_4(q)$ *for some prime power* q.

Finite groups having a (B,N)-pair of rank 1 and satisfying (*) have been determined in [4]. Also Tits [5] has classified groups having a (B,N)-pair of rank $n \geq 3$ and an irreducible Weyl group. Under the additional assumption that the group satisfies (*) we can use Theorem 1 to give an alternative proof of Tits' theorem. In particular, we have

THEOREM 2. *If* G *is a finite simple group having a* (B,N)-*pair and satisfying* (*), *then* G *is a Chevalley group*.

The outline of the proof of Theorem 1 is as follows. By elementary reductions we may assume $G = G_0$, $Z = 1$, and $U = \text{Fit}(B)$. A result of Feit and Higman [2] implies that $|W| = 4, 6, 8, 12$, or 16. The case $|W| = 4$ is treated separately. If $|W| = 6$, the result follows from a result of Higman and McLaughlin [3]. We then show that U is a p-group for some prime p. Using results of Tits and Richen, we construct root subgroups U_r of U, where $r \in \Delta$ and Δ is a root system for W. We also obtain factorizations for U in terms of the root subgroups. These factorizations give precise information about $|U|$ once it is known that $U \cap H = 1$.

If $U \cap U^{s_i} \trianglelefteq U$ for $i = 1,2$, then we establish commutator relations among the groups U_r, $r \in \Delta$. Indeed, under this hypothesis we show that $[U_r,U_s] \leq \Pi U_{ir+js}$ where the product is taken over all $i > 0$ and $j > 0$ such that $ir + js \in \Delta$. These relations are proved by using the above factorization results of Richen and Tits. Unfortunately, showing that $U \cap U^{s_i} \trianglelefteq U$ for $i = 1,2$ is a difficult matter and has to do with the problem of showing that $H \cap U = 1$.

Write $S = \{s_1, s_2\}$ and set $P_i = \langle B, s_i \rangle$ for $i = 1,2$. Then P_i is a maximal parabolic subgroup and P_i has a (B,N)-pair of rank 1 satisfying (*). Thus Theorem C' of [4] gives the structure of R_i/K_i where $K_i = \bigcap_{g \in P_i} B^g$ and $R_i = U^{P_i} K_i$. Indeed, R_i/K_i is isomorphic to $PSL(2,q_i)$, $PSU(3,q_i)$, $Sz(q_i)$, a group of Ree type, a sharply 2-transitive group, or the 2-transitive group of degree 9 and order $3^2 \cdot 2^4$.

A major portion of the paper is involved with showing that the pair $(R_1/K_1, R_2/K_2)$ is of a type occurring in a Chevalley group. For the rank 2 Chevalley groups this pair and $|W|$ determine the group. For example $PSU(4,q)$ is the only simple Chevalley group where $(R_1/K_1, R_2/K_2) = (PSL(2,q), PSL(2,q^2))$ and $|W| = 8$. We begin this pairing by showing that for $i = 1,2$ R_i/K_i is not of Ree type, nor a Frobenius group other than one of the above types, nor the group of order $3^2 \cdot 2^4$. Elimination of the last possibility is difficult. Once this has been done we have $U \cap U^{s_i} \trianglelefteq U$ for $i = 1,2$, $U \cap H = 1$, and the commutator relations mentioned above. Elimination of groups of Ree type and Frobenius groups is then possible. Next we consider the groups P_i acting on $O_p(K_i)$ for $i = 1,2$ and consider the chief factors as representation modules of the groups P_i. By simultaneously considering these modular repre - sentations of P_i and making use of known facts about the representations of $PSL(2,q_i)$, $PSU(3,q_i)$, and $Sz(q_i)$, we complete the pairing of the parabolic subgroup.

After showing that H is abelian, we proceed to the final identification. As remarked earlier, once the pairing of the parabolics is completed, it is clear which Chevalley group is being considered. Thus it suffices to show that the multiplication table

G is uniquely determined. Although this is possible in all cases, we use existing characterization theorems where possible in order to shorten the proof. Thus we determine the centralizer of an involution (or several involutions) in many cases. However, for PSU(5,q) and ${}^2F_4(q)$ the complete multiplication table is determined. This is done by considering the representations of P_i on $O_p(K_i)$ in more detail and making careful choices in the labeling of elements in root subgroups to determine the precise commutator relations holding between root subgroups U_r, U_s for r,s in Δ.

To prove Theorem 2 we set $S = \{s_1,\ldots,s_n\}$ and let $P_{ij} = \langle B,s_i,s_j\rangle$ for $i \neq j \in \{1,\ldots,n\}$. If $K_{ij} = \bigcap_{g \in P_{ij}} B^g$. then P_{ij}/K_{ij} satisfies the hypotheses of Theorem 1 and the tructure of P_{ij}/K_{ij} is determined. If Δ_{ij} is the root system generated by the fundamental roots corresponding to $\{s_i,s_j\}$ then the structure of P_{ij}/K_{ij} is sufficient to determine the commutator relations between root subgroups U_r and U_s for r,s in Δ_{ij}. Then a result of Curtis [1] is applied to complete the proof of Theorem 2.

REFERENCES

[1] C. Curtis, Central extensions of groups of Lie type, Journal für Math. 220 (1965), 174-185.

[2] W. Feit and G. Higman, The nonexistence of certain generalized polygons, J. Algebra 1 (1964), 114-131.

[3] D. G. Higman and J. McLaughlin, Geometric ABA-groups, Illinois J. Math. 5 (1961), 382-397.

[4] W. Kantor and G. Seitz, Some results on 2-transitive groups, Inventiones Math. 13 (1971), 125-142.

[5] J. Tits, Buildings and (B,N)-pairs of spherical type (to appear).

University of Illinois, Chicago Campus

University of Oregon

CHARACTERS OF SYMPLECTIC GROUPS OVER F_2

J. S. Frame and A. Rudvalis

Let S_{2n}, or simply S, denote the 2n-dimensional symplectic group $Sp_{2n}(2)$ over the field with two elements. The group S_{2n} jas (maximal) full orthogonal subgroups $O_{2n}(2,+)$ and $O_{2n}(2,-)$ which we denote by O^+_{2n} and O^-_{2n}. or simply by O^+ and O^-. The groups O are not simple but for $n \geq 3$ have simple commutator subgroups of index two. We let 1 denote the principal (trivial) character of degree one of any group and for groups which have a subgroup of index two we let ε denote the "alternating" character which is equal to +1 for elements inside the subgroup of index two and is equal to -1 for elements outside the subgroup of index two. Since the groups considered here have at most one subgroup of index two the alternating character ε is unambiguously defined (in the cases where it exists). In particular, the groups $O^{\pm}$ always have such a character while the groups S_{2n} have such a character only for $n \leq 2$. If X is any character of a subgroup of S we let X* denote the character of S induced from X.

The permutation representations of S on the cosets of O^+ and O^- are both doubly transitive and in fact we have the following:

LEMMA 1. $(1_{O^-})^* = 1 + X_1$ $\qquad (\varepsilon_{O^-})^* = X_3 + X_4$

$(1_{O^+})^* = 1 + X_2$ $\qquad (\varepsilon_{O^+})^* = X_3 + X_5$

We say that the principal character 1 of S is of <u>level</u> zero while X_1, X_2, X_3, X_4, and X_5 are the characters of <u>level</u> one of S.

The degrees of the characters of level one of S are given by:

LEMMA 2. <u>The degree</u> x_i <u>of the character</u> X_i <u>is as follows</u>:

$$x_1 = (2^n + 1)(2^{n-1}-1) \quad \ldots\ldots\ldots\ldots \quad \binom{10}{01}$$

$$x_2 = (2^n-1)(2^{n-1}+1) \quad \ldots\ldots\ldots\ldots \quad \binom{01}{10}$$

$$x_3 = (2^n-1)(2^n+1)/3 \quad \ldots\ldots\ldots\ldots \ (1/3) \quad \binom{1}{1}$$

$$x_4 = (2^n-1)(2^{n-1}-1)/3 \quad \ldots\ldots\ldots\ldots \ (1/3) \quad \binom{00}{11}$$

$$x_5 = (2^n+1)(2^{n-1}+1)/3 \quad \ldots\ldots\ldots\ldots \ (1/3) \quad \binom{11}{00}$$

The degrees of all characters of S_{2n} appear to be products of an even number of binomials of the form $2^i \pm 1$ $(1 \leq i \leq n)$ times a fraction whose numerator is a power of two and whose denominator is an odd number. Since the binomial factors are (apparently) never repeated and since the fraction is a constant independent of n we have introduced symbols, such as the ones on the right in Lemma 2 above, to denote the degrees of the characters of S. The entries in the top row denote factors of the form 2^i+1 while those in the bottom row denote factors of the form 2^i-1. Thus if there is a 1 in the i-th column of the top (resp. bottom) row of the array, then a factor $2^{n+1-i}+1$ (resp. $2^{n+1-i}-1$) appears in the degree of the character corresponding to the symbol. We observe that the characters of level one have exactly two 1's in the array part of their symbol. In some cases the symbols stand for (degenerate) characters of degree zero or even for the negative of a character. For instance, if $n = 1$, then the characters X_1 and X_4 have degree zero because one of the factors in their symbol is zero, but for $n > 1$ these characters always have degree greater than zero.

The characters of level k for $k > 1$ are defined inductively as those characters of S which are not already of level less than k and which appear as irreducible components in the product of some character of level 1 with some character of level k-1.

LEMMA 3. *A character of level* k *of* S *always has* 2k 1's *in the array part of its symbol*.

In particular, we have:

LEMMA 4. *The characters of level* 2 *of* S *are given by the symbols*:

$(1/9)\ \binom{11}{11}$

$(1/5)\ \binom{11}{11}$

$(2/9)\ \binom{11}{11}$

$(1/9)\ \binom{100}{111}$

$(1/9)\ \binom{111}{100}$

$(1/3)\ \binom{110}{101}$

$(1/3)\ \binom{101}{110}$

$(8/9)\ \binom{101}{101}$

$(8/15)\ \binom{010}{111}$

$(8/15)\ \binom{111}{010}$

$(8/9)\ \binom{110}{011}$

$(8/9)\ \binom{011}{110}$

$(2/15)\ \binom{1000}{1101}$

$(2/15)\ \binom{1101}{1000}$

$(2/9)\ \binom{1100}{1001}$

$(2/9)\ \binom{1001}{1100}$

At this point several observations can be made. First, two distinct symbols can have the same array and differ only in the constant fraction factor. Although we do not have a proof of this it appears that two distinct characters (except possibly algebraically conjugate characters) never have the same symbol. On the other hand, there are some cases in which two distinct symbols denote the same character for one particular value of n but not for any larger values of n. Second, not all possible arrays occur; for instance there is no character (of level two) whose array has 011 in the top row and 101 in the bottom row. As a matter of fact, we do not as yet know (except in a post hoc way) which arrays actually do occur or how to assign the constant fraction to the ones that do occur to complete the symbol. This is a problem which is the center of much of our attention at the present time and we hope to be able to settle this question.

An observation which turns out to be even more interesting than those above is the fact that for each symbol there is a "paired" symbol derived from it by interchanging the top and bottom rows in the array of the given symbol. Some symbols, such as the first three and the eighth in Lemma 4 and also the third in Lemma 2 are left invariant by this involutory mapping and these are called self-paired, as are the characters they denote. Those symbols which are not self-paired yield two types of characters: paired characters, corresponding to paired symbols both of which denote characters of degree greater than zero; and widowed characters, corresponding to paired symbols one of which denotes a character of degree zero. Whether or not a given pair of symbols denotes paired or widowed characters depends on the dimension 2n of S. In fact, if for a certain integer n a symbol denotes a widowed character then for all dimensions > 2n it denotes one member of a pair of characters. Widowed characters seem to appear only at levels greater than or equal to n-1. Furthermore, it appears that it is precisely the widowed characters which are represented by two distinct symbols although we do not as yet have a proof of this.

The pairing described above "respects" multiplication of characters and inducing of characters up from those subgroups of S which are again symplectic and orthogonal groups of lower dimensions.

LEMMA 5. *Let* X *and* Y *be two characters of* S *and let* XY *denote their product (in the* Z*-character ring of* S*). Also, let* X' *and* Y' *denote the pairs of these two characters, by which we mean each irreducible component of* X *(resp.* Y*) is replaced by its pair. Then we have* X'Y' = (XY)'.

LEMMA 6. *Let* T *be a character of* $O^{\pm}$ *and as before let* T* *denote the character of* S *induced from* T. *Then there exists a character* U *of* $O^{\mp}$ *such that* (T*)' = U*.

Caution must be taken in using Lemmas 5 and 6 as certain degeneracies must be taken into account, in particular those arising from the presence of widowed characters. For instance, one of the components of X'Y' might be a widowed character and thus not appear in (XY)'. As a matter of fact, we have examples in which two-sided degeneracies of this type occur, i.e. both XY and X'Y' have irreducible components which are widowed. Nevertheless, such degeneracies should not be considered as too important, for if one is multiplying a character of level i with one of level j then degeneracies can occur only for i + j > n, so that for any product formula one writes down there are only finitely many values of n for which corrections must be introduced.

Implicit in the discussion above is another lemma:

LEMMA 7. *If* X *and* Y *are any two irreducible characters of* S *corresponding to symbols, say* Δ_X *and* Δ_Y*, then there is an expression for* XY *as a non-negative integral linear combination of symbols and this expression is valid for all values of* n *greater than a fixed integer depending only on the levels of* X *and* Y. *This formula remains valid also for smaller values of* n *although one must take into account the fact that some of "components" are of degree less* *than or equal to zero*.

One may also observe that Lemma 6 yields a correspondence between the characters of O^+ and those of O^-, and in a way we do not as yet completely understand, this correspondence may be closely related to the conjecture that one obtains the character table of a twisted Steinberg group defined over a field with q^2 elements by replacing q by -q and changing signs if necessary in the "abstract" character table of the corresponding untwisted group defined over the field with q elements. Nevertheless, we have used Lemma 6 as a starting point in the general theory of characters of the groups $O_{2n}(2,\pm)$.

Another observation which is supported by a continually increasing supply of examples is that the levels defined above seem to be very intimately connected with induction from certain subgroups of S, in a way which we now explain. For the group S_{2n} the subgroups O^+_{2n} and O^-_{2n} as well as their commutator subgroups are defined to be of *depth* one in S. The intersection of O^+_{2n} and O^-_{2n} in S_{2n} is a direct product of Z_2 with $S_{2(n-1)}$, where Z_2 denotes a cyclic group of order two. This intersection, as well as its $S_{2(n-1)}$ subgroup, are defined to be of *depth* two in S. Groups of greater depth in S are then defined inductively, so that a subgroup of *depth* k + 1 in S_{2n} is, except possibly for a factor Z_2, a subgroup of depth k-1 in $S_{2(n-1)}$. Although we do not as yet have a proof of this, examples seem to bear out the following conjecture:

CONJECTURE 8. *If* X *is a character of level* i *in a subgroup of depth* j *in* S, *then the irreducible components of the induced character* X* *are of level at most* i+j.

Strictly speaking the above conjecture makes sense only if i is even, as levels have not been defined here for orthogonal groups. The authors have a working definition of levels for these groups as well although it is somewhat more complicated than in the symplectic case.

We have also found some interesting connections between our levels and pairing for the characters and the modular (Brauer) characters of S. In particular, the trees for most blocks of defect one have their endpoints of relatively low level while the level of the intermediate members of the tree increases in a very regular way toward the center of the tree.

This pattern and even more is illustrated by the blocks of defect one corresponding to the largest prime dividing $2^n+(-1)^n$, i.e., 7 for n = 3 and 17 for n = 4. Thus the 7-chain for $Sp_6(2)$ is:

1		$\binom{0}{0}$	0
27		$\binom{10}{01}$	1
120	(8/9)	$\binom{110}{011}$	2
405	(1/7)	$\binom{111}{111}$	3
512	(256/405)	$\binom{1111}{0110} = (256/315)\binom{0111}{1110}$	3
216	(8/15)	$\binom{111}{010}$	2
15	(1/3)	$\binom{11}{00}$	1

The 17-tree for $Sp_8(2)$ is:

1		$\binom{0}{0}$			0
135		$\binom{01}{10}$			1
4200	(8/9)	$\binom{011}{110}$			2
34560	(256/315)	$\binom{0111}{1110}$			3
65536	$(2^{15}/3^5 5^2 7)$	$\binom{01111}{11110}$	$= (2^{15}/3^4 5\cdot 7\cdot 17)$	$\binom{11111}{01110}$	4
$\overline{42525}$	(17-conjugate pr.)(1/17)	$\binom{1111}{1111}$			4
8960	(256/405)	$\binom{0110}{1111}$			3
1512	(8/15)	$\binom{010}{111}$			2
35	(1/3)	$\binom{00}{11}$			1

The entries in the first column are the character degrees, those in the second column are the symbols, and those in the third column are the levels. One of the two characters of highest level in the block is the Steinberg character of degree 2^{n^2} and since this is a widowed character it has two representations by symbols. The other character of highest level has for its array a solid 2 x n block of 1's for its factor $(1/2^n+(-1)^n)$. The two characters attached to these two are the characters arising from the two representations of the Steinberg character at the lower level and from there to the extremes of the trees there is a regular pattern which differs slightly in the cases where n is odd and those in which it is even. In any case, the behavior of these trees appears to be predictable and the tree for p = 31 for $Sp_{10}(2)$ could be written down without any difficulty.

As in both cases above, the Steinberg character is always widowed and thus has two symbols which are obtained from one another by interchanging the two rows in the array except in the last column. Thus, the two symbols for the Steinberg character are:

$$c\binom{111\ldots111}{011\quad 110} \qquad \text{and } d\binom{011\ldots111}{111\quad 110}$$

the constant factors c and d being 2^{n^2-1}/(product of all the odd factors in the symbol). It is evident at this point that the trees above are built up from symbols which represent Steinberg characters at lower levels.

Also, the two symbols mentioned above which represent the (widowed) Steinberg character are again merely one instance of the pattern for widowed characters, which always have two symbols, one being obtained from the other by interchanging the two rows, except in the last column and then altering the denominator of the constant factor in the appropriate (and indeed predictable) manner.

There is also a beautiful set of patterns for the characters of odd degree which are always 2^{n+1} in number and appear to "grow" out of the characters of odd degree for smaller values of n in an entirely predictable manner which is beautiful to see when entirely written out. (See Appendix)

Many other observations and conjectures appear each time we re-examine the characters of the symplectic and orthogonal groups over the field with two elements, and there are many more facts and observations which we have not mentioned here as they do not appear to be as important as the ones above.

There are strong indications, based on preliminary calculations with orthogonal groups over the field with three elements, that all or almost all of the above facts and observations are just a special case, albeit a beautiful one, of a general theory of characters of orthogonal groups over finite fields. The reason that it is not a special case of a general theory of characters of symplectic groups over finite fields is that for fields of characteristic two, in particular for F_2, symplectic groups of dimension 2n are also orthogonal groups of dimension 2n+1, and for generalizing the theory developed here, it appears to be more natural to view them as orthogonal groups. Taking a somewhat more global view, there is some reason to suspect that something even more fascinating is true, specifically that the theory developed here and the generalization suggested above are again only special cases of a general theory of characters of algebraic groups (in particular families of algebraic groups) over finite fields. The theory developed here might be considered as a theory of characters induced from semi-simple subgroups as opposed to the theory of characters induced from parabolic subgroups as developed by Curtis, Iwahori, Kilmoyer and Benson. If this is indeed the case there are still considerable technical obstacles to a satisfactory theory. For instance in the parabolic case what makes everything work is that the situation is entirely controlled by the geometry (i.e. the Weyl group). In the proposed semi-simple theory the field will also have a significant contribution and it is not clear at the present time whether or not the role of the field can be separated from the geometry of the situation. One fact which makes us hopeful is that in Rudvalis' thesis such a separation was accomplished on a much smaller scale in a closely related problem, one of determining the subdegrees of the action of a symplectic, orthogonal, or unitary group acting on the cosets of a maximal semi-simple subgroup.

AUTHORS' NOTE. Further details, including the character tables of $Sp_6(2)$, $O_8(2,+)$, $O_8(2,-1)$, and $Sp_8(2)$, are available upon request. Address correspondence to either of the authors, preferrably Rudvalis.

APPENDIX

Remarks on Characters of Odd Degree of $Sp_{2n}(2)$

The group $Sp_{2n}(2)$ (for $n > 1$) has 2^{n+1} characters of odd degree, of which 2^n are self-paired, while the remaining 2^n lie in 2^{n-1} "pairs". The self-paired characters of odd degree at level j are 2^{j-1} in number and they are represented by symbols whose array part consists of two rows of length j all of whose entries are 1's. The paired characters of odd degree at level j ($j \geq 3$) are also 2^{j-1} in number and their symbols are ones which represent widowed characters (of even degree) in $Sp_{2j}(2)$, but characters of odd degree in $Sp_{2n}(2)$ for $n > j$. Determining the constant factors associated with the array part of the symbol for the self paired characters is a problem on which we are presently spending considerable time. The table on the following pages illustrates the theory in the cases of $Sp_{2n}(2)$, for n = 1, 2, 3, and 4. We use some notation here which has not been introduced before. Thus, a subscript "w" denotes a widowed character which is represented by two distinct symbols. The expressions in square brackets following the character degrees in the columns headed $Sp_2(2) = \Sigma_3$ and $Sp_4(2) = \Sigma_6$ refer to the Frobenius-Young-Nurnaghan partition symbols identifying the characters of symmetric groups. The subscripts a, b, c, ... in the column headed $Sp_6(2)$ refer to the notation for the characters employed in Frame's paper "The classes and representations of the groups of 27 lines and 28 bitangents," Annali di Mathematica (4)32 (1951), 83-169. Finally, the subscripts ..., i, j, k, ... in the column headed $Sp_8(2)$ refer to the notation used by the authors in their character table of $Sp_8(2)$.

CHARACTERS OF ODD DEGREE FOR $Sp_{2n}(2)$ $(1 \leqq n \leqq 4)$

Symbol	$Sp_2(2) \cong \Sigma_3$	$Sp_4(2) \cong \Sigma_6$	$Sp_6(2)$	$Sp_8(2)$
$\binom{0}{0}$	1 [3]	1 [6]	1	1
(1/3) $\binom{1}{1}$	$1'[1^3]$	$5\ [21^4]$	21_b	85_i
	*********			
(1/3) $\binom{00}{11}$	0_w	$1'[1^6]$	7_a	35_i
(1/3) $\binom{11}{00}$	$2_w[21]$	$5\ [2^3]$	15_a	51_i
$\binom{10}{01}$	0_w	5 [51]	27_a	119_i
$\binom{01}{10}$	$2_w[21]$	9 [42]	35_b	135_i
(1/9) $\binom{11}{11}$		$5\ [3^2]$	105_b	1785_i
(1/5) $\binom{11}{11}$		$9[2^2 1^2]$	189_b	3213_j
		*******		
(1/9) $\binom{100}{111}$		0_w	21_a	595_i
(1/9) $\binom{111}{100}$		$10_w[31^3]$	105_c	1275_i
(1/3) $\binom{110}{101}$		0_w	105_a	2295_i
(1/3) $\binom{101}{110}$		$10_w[31^3]$	189_c	2975_j

Symbol	$Sp_6(2)$	$Sp_8(2)$
(1/81) $\binom{111}{111}$	35_a	2975_i
(1/15) $\binom{111}{111}$	189_a	16065_h
(1/9) $\binom{111}{111}$	315_a	26775_j
(1/7) $\binom{111}{111}$	405_a	34425_i
	****	
(1/27) $\binom{1100}{1111}$	0_w	1785_j
(1/27) $\binom{1111}{1100}$	$(210_b)_w$	8925_j
(1/9) $\binom{1110}{1101}$	0_w	8925_i
(1/9) $\binom{1101}{1110}$	$(210_b)_w$	16065_i
(1/15) $\binom{1100}{1111}$	0_w	3213_i
(1/15) $\binom{1111}{1100}$	$(378_a)_w$	16065_j
(1/5) $\binom{1110}{1101}$	0_w	16065_g
(1/5) $\binom{1101}{1110}$	$(378_a)_w$	28917_i

CHARACTERS OF ODD DEGREE FOR $Sp_{2n}(2)$ $(1 \leqq n \leqq 4)$ continued

Symbol	$Sp_8(2)$
$(1/1215)\binom{1111}{1111}$	595_j
$(1/75)\binom{1111}{1111}$	9639_i
$(1/45)\binom{1111}{1111}$	16065_k
$(1/27)\binom{1111}{1111}$	26775_i
$(1/21)\binom{1111}{1111}$	34425_j
$(1/17)\binom{1111}{1111}$	42525_i
$(1/17)\binom{1111}{1111}$	$42525_j = \overline{42525}_i$
$(1/15)\binom{1111}{1111}$	48195_i
	*******

Michigan State University

University of Massachusetts

STRONGLY CLOSED ABELIAN 2-SUBGROUPS OF FINITE GROUPS

David M. Goldschmidt

Recall that for groups $A \subseteq T \subseteq G$, we say that A _is strongly closed in_ T _with respect to_ G _if_ $a \in A$, $g \in G$ _and_ $a^g \in T$ _implies_ $a^g \in A$.

For the past several years, I have been studying various special cases of the above situation, with $T \in Syl_2(G)$ and A abelian. These conditions arise whenever a 2-local subgroup $N_G(H)$ controls the fusion of its 2-elements, for it is easy to see that in this case, Z(H) is strongly closed in any Sylow 2-subgroup containing it. Conversely, Glauberman has shown that if A is abelian and strongly closed in a Sylow 2-subgroup of G, then $N_G(A)$ controls the fusion of its 2-elements.

There seems to be good reason to believe that if A is abelian and strongly closed in a Sylow 2-subgroup of G, then the normal closure of A in G has composition factors of known type. I hope to have a proof of this result shortly.

Note that any of the following hypotheses implies the existence of a strongly closed abelian 2-subgroup:

a) Sylow 2-subgroups of G are abelian.

b) Sylow 2-subgroups of G are of type $S_z(q)$ or $U_3(q)$.

c) A Sylow 2-subgroup contains an isolated involution.

d) G has a weakly embedded 2-local subgroup.

e) The weak closure of a central involution in its centralizer is abelian.

Simple groups satisfying the above hypotheses have been classified by various authors. Thus, in addition to providing new information, the proposed classification generalizes the above results, and should give independent proofs for a), b), d) and possibly e).

University of California, Berkeley

FINITE GROUPS OF SECTIONAL 2-RANK AT MOST 4

Daniel Gorenstein and Koichiro Harada

A number of people have studied the possible 2-groups of low rank that can occur as a Sylow 2-subgroup of a simple group, using primarily fusion methods together with some character theory. Probably the first person was Thompson in his work in the N-group paper for the case that $SCN_3(2)$ was empty. MacWilliams, in her Ph.D. thesis under Thompson, then considered the general case of $SCN_3(2)$ empty. After that, Alperin completely analyzed the rank 2 case and studied the rank 3 case extensively. In addition, both Fong and Harada obtained a number of results concerning Sylow 2-subgroups of low order.

Several years ago, in conversations with Alperin and I think once with Thompson, I expressed the view that this entire effort would be greatly facilitated if one were in an inductive situation in which the composition factors of the proper subgroups were known. I felt strongly that the various analyses so far carried out were accomplished "with one hand tied behind one's back", as the saying goes. My own views about this had developed out of my work on groups with dihedral and quasi-dihedral Sylow 2-subgroups, in which detailed properties of the known simple groups with such Sylow 2-subgroups played a fundamental role throughout the analysis as a consequence of the fact that these classification theorems were proved inductively.

So a little over a year ago when Harada came to Rutgers for a few months, it was reasonable for me to suggest to him that we attack the sectional 2-rank at most 4 problem -- that is, every section of G has 2-rank at most 4; equivalently, every 2-subgroup of G can be generated by at most 4 elements. By a result of MacWilliams, this included all groups in which $SCN_3(2)$ was empty.

Here then was a nice inductive situation. We would know all the possible composition factors of the proper subgroups of G in advance and so our task was obviously going to be much easier than MacWilliams'. All we had to do was to pin down the possibilities for a Sylow 2-subgroup of G and that would be it! For Harada and I had previously characterized almost all of the known simple groups of 2-ranks 3 and 4 and sectional 2-rank at most 4 by their Sylow 2-subgroups. Apart from those with abelian Sylow 2-subgroups, the remaining ones have since been similarly characterized by David Mason. The ones with abelian Sylow 2-subgroups were, of course, known; and Alperin's result in the rank 2 case plus several classification theorems took care of all the simple groups of 2-rank at most 2.

There it was--a fairly easy problem for Harada and me to knock off with just a little effort. First of all, in the case that all 2-local subgroups were 2-constrained, we would kill off the cores, using my theorem with John Walter about balanced groups (when $SCN_3(2)$ was nonempty). So how hard could it then be from that point to force the structure of a Sylow 2-subgroup of G? Well, this would bring us to the non-constrained case. Clearly then, having a component sitting inside of a 2-local subgroup, must surely mean that we could practically read off a Sylow 2-subgroup of G at sight!

There you have Gorenstein's "game plan"! So what was Phase I? First, we dispose of the "trivial" case, in which all 2-locals are solvable. I failed to become suspicious when this case didn't succumb so easily. In the end, Harada and I had to make use of the methods of Section 14 of the N-group paper to achieve a first approximation of the possible Sylow 2-subgroups of G. At that point Harada performed the kind of delicate fusion operation of which he is a master to determine the exact possibilities for the Sylow 2-subgroup of G.

Being a natural optimist, I refused to face the full implications of our analysis, but busied myself writing up our results, while Harada turned his attention to Phase II -- all 2-locals constrained and at least one nonsolvable. I was not prepared at all for the analysis he carried out. First of all, he completely discarded my game plan by ignoring the idea of killing the cores (which at least we did have to do in the solvable case). Instead he treated the problem first as a purely 2-local question and then as a purely fusion-theoretic 2-group problem. The total argument constituted the most delicate and imaginative analysis of this type that I have ever seen.

However, still undaunted, I remained convinced that at least the non-constrained case would be easy. But the cruellest insult of all awaited me, for we were forced to expand Phase III into Phases II, IV, V, and VI! Obviously with such a game plan, it was time to throw out the administration. But fortunately Harada had never been as naive as I about the difficulties involved. Five years of work had taught him to respect all 2-group problems. So while I was still expounding the administration's view, Harada was quietly preparing sound fallback positions. Gradually there emerged a fairly elaborate subdivision of the non-constrained case, each of which we have now been able to resolve successfully.

What went wrong then? Well, it turns out that the existence of a component, particularly one like $L_2(q)$, $L_3(q)$, or $SL(2,q)$, q odd, in the centralizer of an involution doesn't really tell you at the outset very much about the possible structures of a Sylow 2-subgroup of G, even when the involution in question is a central one. Basically to get anywhere, we were forced to study the normalizers in G of the elementary 2-subgroups of maximal rank. However, it took an enormous amount of work to translate the initial information into effective results about these normalizers. Eventually these arguments produced an approximate structure of both a Sylow 2-subgroup and a non-constrained centralizer of an involution of G. Now Harada was able to finish off the problem by fusion arguments. However, even then, the number of individual cases that had to be treated separately at the end was surprisingly large.

Thus this "simple" problem I suggested to Harada has transformed itself into another 500-page monster with 6 heads. Although in one sense, the methods used can be regarded to be standard, there is no question in my mind that at the very least we have developed important procedures that will be useful in dealing with other low 2-rank problems and which should be capable of considerable generalization. Thus it will profit someone to read our paper carefully. For myself, this whole experience has constituted quite an education. I assure you that I shall never again underestimate the complexity of a 2-group problem in any situation!

I should like now to state our main result together with some corollaries and give a brief outline of the six individual parts of the paper along with a few comments. I should say first, however, that Harada and I are now in the process of writing everything up in final form and so this outline should be considered as somewhat tentative.

As I remarked before, Part I, which deals with the case of solvable 2-locals, utilizes some important ideas of the N-group paper as well as a result of Janko and Thompson to take care of the $SCN_3(2)$ empty case. Moreover, we note that the condition that 2-locals be solvable is itself inductive to all sections and so this case can be considered to be an independent inductive problem.

The analysis of Part II hinges on Theorem B, which is a purely local result giving the structure of a nonsolvable 2-constrained group of sectional 2-rank at most 4. My own feeling is that of all the individual results in our paper, this one will be the most difficult to extend to, say, the 2-rank 4 case. I suspect that one may require a more global type of analysis, similar in spirit to that of Sections 13, 14, and 15 of the N-group paper, in order to pin down the possibilities for $O_2(X)$.

On the basis of this local Theorem B, Harada was able to complete the case of 2-constrained 2-locals and, in addition, was able to handle two further situations which turned out to be very important for the later subdivision of the non-constrained case. These are Theorems C and D of Part II.

Beginning in Part III, we bring our full induction hypothesis into play, so that henceforth the nonsolvable composition factors of the proper subgroups of G are assumed to satisfy the conclusion of the Main Theorem. In theorem A, we then determine the possible components that can occur in a non-constrained centralizer of an involution x of G. This is primarily a problem of Schur multipliers. In the corollary of Theorem A we determine the possible 2-layers L_x of $C_G(x)/O(C_G(x))$. The principal result of Part III is the determination of G in the case that L_x has sectional 2-rank exactly 4 with each of its components nonsimple for some involution x of G.

Here is where a large number of individual cases must be considered. Some of them are quite delicate--particularly the $SL(2,q_1) * SL(2,q_2)$, q_1, q_2 odd, case.

In parts IV, V, and VI, we begin to bring our previous results into the picture along with a few classification theorems. We thus successively define the notion of $\mathcal{R}_1$-group, an $\mathcal{R}_2$-group, and an $\mathcal{R}_3$-group and show in turn that it suffices to establish the Main Theorem, first, for $\mathcal{R}_1$-groups, then for $\mathcal{R}_2$-groups, and finally for $\mathcal{R}_3$-groups. Parts IV and V together treat the case in which the centralizer of some central involution x is not 2-constrained. Since the case that L_x has sectional 2-rank 4 with all components nonsimple and the case that a Sylow 2-subgroup of G is abelian or is of 2-rank 2 have been excluded in the definition of an $\mathcal{R}_1$-group, there exists only a single simple group which satisfies the given conditions--namely, $D_4^2(3)$, in which case correspondingly $L_x \cong SL(2,27)$.

Finally in Part VI we are left with the case that the centralizer of every central involution is solvable and the centralizer of some noncentral involution is not 2-constrained. Here again there is only one simple group which fits the given conditions--namely, M_{12}, in which case correspondingly $L_x \cong L_2(5)$. Pinpointing the Sylow 2-subgroup of G in this case is enormously complex, requiring over a 100 pages of delicate analysis. I find it very remarkable indeed that out of the long analyses that occur in Parts IV, V, and VI, only two groups emerge--$D_4^2(3)$ and M_{12}--both of which have the same Sylow 2-subgroups!

This is as much as I can say here about the proof. The statement of the Main Theorem, some corollaries, and the principal results of each part of the paper are given below.

MAIN THEOREM. *If G is a finite simple group of sectional 2-rank at most 4, then G is isomorphic to one of the groups in the following list:*

(I). *Odd characteristic:*

$L_2(q)$, $L_3(q)$, $U_3(q)$, $G_2(q)$, $D_4^2(q)$, $PSp(4,q)$, $G_2^1(q)$, q *an odd power of* 3, $L_4(q)$, $q \not\equiv 1 \pmod 8$, *or* $U_4(q)$, $q \not\equiv 7 \pmod 8$;

(II). *Even characteristic:*

$L_2(8)$, $L_2(16)$, $L_3(4)$, $U_3(4)$, *or* $Sz(8)$;

(III). *Alternating:*

A_7, A_8, A_9, A_{10}, *or* A_{11};

(IV). *Sporadic:*

M_{11}, M_{12}, M_{22}, M_{23}, J_1, J_2, J_3, M^c, *or* L.

COROLLARY A. *If G is a simple group in which* $SCN_3(2)$ *is empty, then G is isomorphic to one of the groups listed in the Main Theorem.*

COROLLARY B. *If G is a quasisimple group of 2-rank 2 with* $O(G) = 1$, *then either G is simple or* $G \cong Sp(4,q)$, *q odd.*

With the aid of Theorem B os Part II, we can prove the following additional result:

COROLLARY C. *If G is a nonsolvable fusion-simple group, of sectional 2-rank at most 4, then one of the following holds:*

(i) G' *is simple;*

(ii) G' *is the direct product of two simple groups of sectional 2-rank 2;*

(iii) G' *is the direct product of a simple group of sectional 2-rank 2 and* $Z_{2^n} \times Z_{2^n}$ *for some* n; *or*

(iv) G' *is a nontrivial extension of* E_8 *or* E_{16} by A_5, A_6, A_7, *or* $L_3(2)$.

Part I

THEOREM A. *If* G *is a simple group of sectional* 2-*rank at most* 4 *in which every* 2-*local subgroup is solvable*, *then* G *is isomorphic to one of the following groups*:

$L_2(q)$, q *odd*, $L_3(3)$, $U_3(3)$, $L_2(8)$, $L_2(16)$, $U_3(4)$, $Sz(8)$, A_7, *or* M_{11}.

Part II

THEOREM A. *If* G *is a simple group of sectional* 2-*rank at most* 4 *in which every* 2-*local subgroup is* 2-*constrained*, *then* G *is isomorphic to one of the following groups*:

(I). $L_2(q)$, q *odd*, $L_3(3)$, $U_3(3)$, $U_4(3)$, $PSp(4,3)$, *or* $G_2(3)$;

(II). $L_2(8)$, $L_2(16)$, $L_3(4)$, $U_3(4)$, *or* $Sz(8)$;

(III). A_7 or A_8;

(IV). M_{11}, M_{22}, M_{23}, *or* J_3.

THEOREM B. *If* X *is a nonsolvable* 2-*constrained group of sectional* 2-*rank at most* 4 *with* $O(X) = 1$, *then one of the following holds*:

(I). $X/O_2(X) = L_3(2)$ *and* $O_2(X) = E_8$, E_{16}, *or* $Z_4 \times E_8$;

(II). $X/O_2(X)$ *contains a subgroup isomorphic to* A_5 *and* $O_2(X) = E_{16}$ *or* $Q_8 * D_8$.

THEOREM C. *If* G *is a simple group of sectional* 2-*rank at most* 4 *in which the centralizer of every central involution is solvable and some* 2-*local subgroup of* G *is nonsolvable and* 2-*constrained*, *then* G *is isomorphic to one of the following groups*:

$L_4(3)$, $U_4(3)$, $PSp(4,3)$, $G_2(3)$, $L_3(4)$, A_8, A_9, A_{10}, A_{11}, *or* M_{22}.

THEOREM D. *If* G *is a simple group of sectional* 2-*rank at most* 4 *in which the centralizer of some involution is nonsolvable and* 2-*constrained, then* G *is isomorphic to* J_2, J_3, *or* M_{23}.

Part III

For any group H, L(H) denotes the unique largest semi-simple normal subgroup of H. For any group G, $\mathcal{L}(G)$ denotes the set of quasisimple components of the groups $L(C_G(x)/O(C_G(x)))$ as x ranges over the involutions of G. Similarly we define $\mathcal{L}_c(G)$, where x is now restricted to the central involutions of G. For brevity, we also write L_x for $L(C_G(x)/O(C_G(x)))$.

THEOREM A. *If* G *is a simple group of sectional* 2-*rank at most* 4 *in which the nonsolvable composition factors of the proper subgroups of* G *satisfy the conclusion of the Main Theorem, then every element of* $\mathcal{L}(G)$ *is isomorphic to one of the following groups*:

(I). *Simple Groups*:

$L_2(q)$, $L_3(q)$, $U_3(q)$, q *odd*, $G_2^1(q)$, q *an odd power of* 3, $K_2(8)$, $Sz(8)$, A_7, M_{11}, *or* J_1;

(II). *Nonsimple groups*:

$SL(2,q)$, $Sp(4,q)$, q *odd*, $SL(4,q)$, $q \equiv 3 \pmod 4$, $SU(4,q)$, $q \equiv 1 \pmod 4$, $\hat{Sz}(8)$, $\hat{A}_n$, $7 \leq n \leq 11$, or $\hat{M}_{12}$.

Here $\hat{Sz}(8)$, $\hat{A}_n$, and $\hat{M}_{12}$ denote the unique perfect central extension of $Sz(8)$, A_n, and M_{12}, respectively by Z_2.

COROLLARY A. *For each involution* x *of* G, *either* L_x *is quasisimple or* L_x *is a central product of two components* L_1, L_2 *such that*

(i) $L_1 \cong SL(2,q)$, q odd, or $\hat{A}_7$; and

(ii) $L_2 \cong L_2(q)$, $SL(2,q)$, q odd, $L_3(q)$, $q \equiv -1 \pmod 4$, $U_3(q)$, $q \equiv 1 \pmod 4$, A_7, $\hat{A}_7$, or M_{11}.

THEOREM B. *Let G be a simple group of sectional 2-rank at most 4 in which the nonsolvable composition factors of the proper subgroups of G satisfy the conclusion of the Main Theorem. If for some involution x of G, L_x has sectional 2-rank 4 and each of its components is nonsimple, then G is isomorphic to one of the following groups*:

(I). $G_2(q)$, ${}^2D_4(q)$, $PSp(4,q)$, q odd, $q \geq 5$;

(II). $L_4(q)$, $q \not\equiv 1 \pmod 8$, $q \geq 5$, or $U_4(q)$, $q \not\equiv 7 \pmod 8$;

(III). M^c or L.

Part IV

We now define an $\mathcal{R}_1$-*group* to be any group G which satisfies the following conditions:

(a) G has 2-rank at least 3, sectional 2-rank at most 4, and nonabelian Sylow 2-subgroups;

(b) The nonsolvable composition factors of the proper subgroups of G satisfy the conclusion of the Main Theorem;

(c) $\mathcal{L}(G)$ is nonempty;

(d) The centralizer of every involution of G is either solvable or non 2-constrained;

(e) For no involution x of G is L_x of sectional 2-rank 4 with each of its components nonsimple.

On the basis of our previous results, together with the classification of simple groups with abelian Sylow 2-subgroups and simple groups of 2-rank 2, it suffices to show that the Main Theorem holds for simple R_1-groups.

THEOREM A. *If* G *is a simple* R_1*-group in which* $L_c(G)$ *is nonempty and each element of* $L_c(G)$ *is nonsimple, then* G *is isomorphic to* $D_4^2(3)$.

Part V

An R_1-group G will be called an R_2*-group* if either $L_c(G)$ is empty or some element of $L_c(G)$ is simple.

By our previous results, it suffices to establish the Main Theorem for simple R_2-groups.

THEOREM A. *If* G *is a simple* R_2*-group, then the centralizer of every central involution is solvable* (*equivalently* $L_c(G)$ *is empty*).

Part IV

An R_2-group will be called an R_3*-group* provided

(a) The cen ralizer of every central involution is solvable;

and (b) Every 2-local subgroup is either solvable or non 2-constrained.

By our previous results, it suffices to establish the Main Theorem for simple R_3-groups.

THEOREM A. *If* G *is a simple* R_3*-group, then* G *is isomorphic to* M_{12}.

Rutgers University

Ohio State University

AUTOMORPHISMS OF EXTRA SPECIAL GROUPS AND NONVANISHING DEGREE 2 COHOMOLOGY

Robert L. Griess, Jr.

We announce some results about group extensions. Proofs will appear in a paper of the same title.

THEOREM 1. *Let* E *be an extra-special group of order* 2^{2n+1}, $n \geq 1$. *We identify* Out(E) *with the relevant orthogonal group* $O^{\varepsilon}(2n,2)$, $\varepsilon = \pm$. *Consider the exact sequences of groups*

$$(*)\quad 1 \to \mathrm{Inn}(E) \to \mathrm{Aut}(E) \overset{\pi}{\to} O^{\varepsilon}(2n,2) \to 1,$$

$$1 \to \mathrm{Inn}(E) \to A(E) \overset{\pi}{\to} \Omega^{\varepsilon}(2n,2) \to 1,$$

where A(E) *is the preimage of* $\Omega^{\varepsilon}(2n,2)$ *in* Aut(E) *under* π

(a) *If* E *is of positive type, the sequences of* (*) *are nonsplit when* $n \geq 3$, *split when* $n = 1,2$.

(b) *If* E *is of negative type, the sequences of* (*) *are nonsplit when* $n \geq 3$, *split when* $n - 1,2$.

COROLLARY 1. *The second cohomology groups* $H^2(O^{\pm}(2n,2),V)$ $H^2(\Omega^{\pm}(2n,2),V)$ *are nonzero for* $n \geq 3$, *where* V *is the usual* 2n-*dimensional space on which* $O^{\pm}(2n,2)$ *acts. Also,* $H^2(O^{+}(4,2),V) = 0$ *and* $H^2(\Omega^{+}(4,2),V) = 0$.

COROLLARY 2. *Let* E *be an extra special group of order* 2^{2n+1} *and let* Y *be cyclic of order* 4. *Then* $\mathrm{Out}(E\circ Y) \cong Z_2 \times Sp(2n,2)$. *The extensions*

$$1 \to \mathrm{Inn}(E\circ Y) \to \mathrm{Aut}(E\circ Y) \overset{\pi}{\to} Z_2 \times Sp(2n,2) \to 1,$$

$$1 \to \mathrm{Inn}(E\circ Y) \to A(E\circ Y) \overset{\pi}{\to} Sp(2n,2) \to 1,$$

where $A(E \circ Y)$ *is the centralizer in* $\mathrm{Aut}(E \circ Y)$ *of* Y, *are nonsplit for* $n \geq 3$. *Consequently, the second cohomology group* $H^2(Sp(2n,2),V)$ *is nonzero for* $n \geq 3$, *where* V *is the standard* $2n$-*dimensional module for* $Sp(2n,2)$.

THEOREM 2. $H^2(O^-(4,2),V) = 0$ *and* $H^2(\Omega^-(4,2),V) = 0$, *where* V *is the usual* 4-*dimensional module on which* $O^-(4,2)$ *acts*.

THEOREM 3. (J. Mc Laughlin) $H^2(Sp(4,2),V) \neq 0$, *where* V *is the usual* 4-*dimensional module. More precisely, the cup product*

$$H^1(Sp(4,2),Z_2) \times H^1(Sp(4,2),V) \to H^2(Sp(4,2),V)$$

is not the zero pairing.

THEOREM 4. $H^2(Sp(4,2)',V) = 0$, V *the standard module*.

THEOREM 5. (a) *Let* E *be an extra special group of order* 2^{2n+1} $n \geq 1$, *type* $\varepsilon = \pm$. *There is a group* H *having the properties*

$$O_2(H) \cong E, \quad Z(H) = Z(O_2(H))$$

$$H/O_2(H) \cong O^{\varepsilon}(2n,2), \quad H/Z(H) \cong \mathrm{Aut}(E)$$

and H *has a faithful, ordinary, irreducible complex representation of degree* 2^n.

(b) *Let* E *be as above and let* $Y \cong Z_4$. *There is a group* H_0 *having the properties*

$$O_2(H_0) \cong E \circ Y \quad Z(H_0) = Z(E)$$

$$H_0/O_2(H_0) \cong Sp(2n,2), \quad H_0/Z(H_0) \cong A(E \circ Y)$$

and H_0 *has two faithful, ordinary irreducible complex representations of degree* 2^n. *These are interchanged by the action of* $\mathrm{Aut}(E \circ Y)$ *on* $A(E \circ Y)$ *and by complex conjugation*.

(c) *Let* F *be a splitting field for* E *or* $E \circ Z_4$ (e.g. $F = \mathbb{Q}$, *the rationals, will do for* E_n, *and* $\mathbb{Q}(\sqrt{-1})$ *will do for* T_n *or* $E \circ Z_4$.

(i) *If* $W = H$ *or* H_0, *then the* 2^n*-dimensional representation, when restricted to* W', *may be written in* F

(ii) *If* $W_0 = H$, $H'E$, H_0 *or* $H_0'E$ *then every faithful irreducible character of* χ *of* W_0 *has the form* $\chi = \xi\eta$, *where* E *is contained in the kernel of* ξ *and* η *is the character of a* 2^n*-dimensional representation from* (a) *if* $W_0 = H$ *or* $H'E$, *or from* (b) *if* $W_0 = H_0$ *or* H_0'.

COROLLARY 3. *The exact sequences of Corollary* 2 *are nonsplit for* $n = 2$.

Note that Corollary 3 implies the first assertion of Theorem 3.

It seems worthwhile to point out that the arguments used to prove Theorem 1, the case $n \geq 4$, actually prove a stronger result. Namely, we get a criterion for a subgroup of $O^\varepsilon(2n,2)$ to have a nonsplit extension over the standard 2n-dimensional module.

Let E be an extra special 2-group of order 2^{2n+1}, $n \geq 1$. If we identify E' with GF(2), the vector space E/E' of dimension 2n over GF(2) has a quadratic form q taking values in E'. It is defined by

$$q(xE') = x^2 \in E'$$

and the associated bilinear form is just commutation:

$$b(xE',yE') = q(xyE')q(xE')q(yE') = [x,y].$$

Clearly, Aut(E) preserves these forms. Since Inn(E) coincides with the group of automorphisms trivial on E/E', we have an embedding

$$\mathrm{Out}(E) \to O^\varepsilon(2n,2).$$

Here, $\varepsilon = \pm$, and $\varepsilon = +$ if E can be expressed as the central product of n dihedral groups, $\varepsilon = -$ otherwise (i.e., if Q_8 is required).

Actually, a generator and relations argument shows that every orthogonal transformation on E/E' can be lifted to E. Thus, we have

$$\mathrm{Out}(E) \cong O^{\varepsilon}(2n,2).$$

From now on, we identify these two groups. Also, we identify E/E' with Inn(E), thus making Inn(E) the standard module for Out(E).

Let G be a subgroup of Out(E). Consider (*), the extension of G by Inn(E) induced by Aut(E):

$$\begin{array}{cccccccccc} & 1 & \to & \mathrm{Inn}(E) & \to & \mathrm{Aut}(E) & \to & \mathrm{Out}(E) & \to & 1 \\ & & & \| & & \uparrow & & \uparrow & & \\ (*) & 1 & \to & \mathrm{Inn}(E) & \to & B & \to & G & \to & 1 \end{array}$$

THEOREM 6. *Let W be a subspace of Inn(E), $|W| = 4$, so that the nontrivial elements of W consist of two singular vectors x,y and one nonsingular vector z. Suppose G has a subgroup K satisfying*

(a) *K fixes the vector z;*

(b) *K contains an involution t with $x^t = y$, $y^t = x$, $z^t = z$;*

(c) *K has no subgroup of index 2.*

Then, (*) is a nonsplit extension.

PROOF. Let W* be the preimage of W in E. Then $W^* \cong D_8$. Assume (*) is split. Then, there is a complement to Inn(E) in B, which we may as well identify with G. Now, $t \in G$ effects an outer automorphism of order 2 on W*. The structure of Aut(W*) implies that t inverts the maximal cyclic subgroup U of W*. But K acts on U, and since K has no subgroup of index 2, K must centralize U. This gives a contradiction, since $t \in K$. Therefore, the complement does not exist, and the theorem is proven.

COROLLARY 4. *Let* $n \geq 4$. *Then, the extension of* Out(E) *or* Out(E') $\cong \Omega^{\varepsilon}(2n,2)$ *by* Inn(E) *is nonsplit. Consequently, if* V *denotes the standard* 2n-*dimensional module for the orthogonal group over* GF(2), *then* $H^2(O^{\varepsilon}(2n,2),V)$ *and* $H^2(\Omega^{\varepsilon}(2n,2),V)$ *are nonzero for* $n \geq 4$.

PROOF. Let G = Out(E) or Out(E)'. Choose any W as in the Theorem 6 and let K be the stabilizer of the nonsingular vector of W in Out(E)'; we have $K \cong O(2n-1,2) \cong Sp(2n-2,2)$. Since $n \geq 4$, K is simple. One can easily find an involution t with the required properties. The Theorem 6 now implies the Corollary.

REMARK. The conclusion of the Corollary holds even for n = 3 (see the author's paper referred to at the beginning of this article). The above methods do not work, however, since $Sp(4,2) \cong \Sigma_6$ has a subgroup of index 2.

EXAMPLE. It may be easy to check the hypotheses of the Theorem 6 in some cases. Let G be Conway's group .1. Let Λ be the Leech lattice, and set $V = \Lambda/2\Lambda$. The quadratic form on Λ induces a GF(2)-valued quadratic form on V which is nondegenerate. Also, G preserves the form on V. Consider a triangle in Λ of type 223. We take for W the union of $0 \in V$ and the images in V of the edges of this triangle. Let K* be the stabilizer in .0 of the subgroup of Λ generated by the edge of type 3; $K^* \cong .3 \times Z(.0)$. Let K be the image of K* in G. It is not difficult to find an involution t* of .0 which switches the two edges of type 2 in our triangle. Finally, set t equal to the image of t* in G. Since K is simple, all parts of the hypothesis of our Theorem are satisfied. It follows that there is a nonsplit extension of G by V (we have embedded G in Out(E), E extra special of order 2^{25}, and identified V with E/E' to get this). Of course, we also get $H^2(.1, \Lambda/2\Lambda) \neq 0$.

University of Michigan

CHARACTERIZATIONS OF SOME FINITE SIMPLE CHEVALLEY GROUPS BY CENTRALIZERS OF INVOLUTIONS

Morton E. Harris

The purpose of this note is to announce the following four theorems which have not as yet appeared in print ([2],[3],[4],[5]) and to describe their proofs:

THEOREM 1. *Let* G *be a finite group having a central involution* j *such that*:

a) $O(C_G(j)) = \{1\}$;

b) $C_G(j)$ *contains a normal subgroup* $\mathcal{C}$ *of odd index* ρ *such that* $\mathcal{C}$ *is isomorphic to the centralizer of a central involution in the finite simple group* $G_2(q)$ *with* $q = 2^n$. *Then at least one of the following hold:*

1) $O(G) \neq \{1\}$;

2) $O(G) = \{1\}$ *and there exists a nontrivial subgroup* E *of* $Z(\mathcal{C})$ *such that* $E \triangleleft G$;

3) $G = O(G)C_G(j)$;

4) G *is isomorphic to an extension of* $G_2(q)$ *by a field automorphism of order* ρ.

THEOREM 2. *Replace* $G_2(q)$ by $D_4^2(q^3)$ *in the statement of Theorem* 1.

Before stating Theorem 3, we note that an extremal involution of PSp(8,q), q odd, is an involution in PSp(8,q) that comes from an involution of Sp(8,q) whose subspace of fixed vectors has dimension 6.

THEOREM 3. _Let_ G _be a finite group with an involution_ j _such that:_

a) $O(C_G(j)) = \{1\}$;

b) $C_G(j)$ _contains a normal subgroup_ $\mathcal{C}$ _of odd index_ ρ _such that_ $\mathcal{C}$ _is isomorphic to the centralizer of an extremal involution in_ PSp(8,q) _with_ q _odd. Then either_

(i) $G = O(G)C_G(j)$ _or_

(ii) G _is isomorphic to an extension of_ PSp(8,q) _by a field of order_ ρ.

THEOREM 4. _Let_ G _be a finite group with an involution_ j _such that:_

a) $O(C_G(j)) = \{1\}$;

b) $C_G(j)$ _contains a normal subgroup_ $\mathcal{C}$ _of odd index_ ρ _such that_ p _is isomorphic to_ Spin(2n+1,q) _with_ q _odd and_ $n \geq 4$.

Then either

(i) $G = O(G)C_G(j)$ _or_

(ii) $n = 4$ _and_ G _is isomorphic to an extension of_ $F_4(q)$ _by a field automorphism of order_ ρ.

Theorems 1 and 2 generalize [7, Theorem] and [8, Theorem] respectively.

Theorem 3 (with $\rho = 1$) yields a characterization of PSp(8,q), q odd, and thereby completes the work of [10], [11], and [12]. Also, Theorem 4 (with $\rho = 1$) characterizes the finite simple groups $F_4(q)$. q odd.

Clearly Theorem 4 poses the question: do Spin (5,q) and/or Spin (7,q), with q odd, occur as the centralizer of an involution in a fusion simple group? Since Spin (5,q) ≅ Sp(4,q), it follows from [12,(1A)] and the Krull-Schmidt Theorem that Spin(5,q) is excluded. The case of Spin(7,q) is being studied by R. Solomon and it is very likely that this case will also be excluded.

In all four theorems, the "extension of $\mathcal{C}$ by a group of order p" is handled by the methods of [6], so we shall assume that p = 1 in describing the proofs of our theorems.

Since the proofs of Theorems 1 and 2 are similar, we shall only describe the proof of Theorem 1.

PROOF OF THEOREM 1. Assume the hypotheses of Theorem 1 and set $D = O_2(\mathcal{C})$ and $Z = Z(\mathcal{C})$. Since j is central a Sylow 2-subgroup of $\mathcal{C}$ is a Sylow 2-subgroup of G. Then, from the structure of $\mathcal{C}$, D contains a subgroup M that is characteristic in any 2-subgroup of G containing M. Be refining the methods of [7], we show:

(i) if $N_G(M)$ is not 2-closed, then G satisfies conclusion (4) of the theorem;

(ii) if $N_G(M)$ is 2-closed, then no involution of $\mathcal{C}$-Z is conjugate in G to an element of Z.

Let k denote the G-conjugacy class of the involution j and set $\Theta = \langle k \cap \mathcal{C} \rangle$ and $H = \langle k \rangle$. Then $\Theta \subseteq Z$, $H \trianglelefteq G$ and $Z(H) \trianglelefteq G$. Since we may assume that $O(G) = \{1\}$, Z(H) is a 2-group. Also $j \in H$ so that $Z(H) \leq D$.

If $Z(H) \neq \{1\}$, then $E = Z(H) \cap Z \neq \{1\}$ and $E \trianglelefteq G$ so that conclusion (2) of the theorem holds. Thus we assume that $Z(H) = \{1\}$. Applying [6], it follows that $H = X_1 \times X_2 \times \ldots \times X_u$ where X_i is a simple group of Bender type.

But $D = O_2(\mathcal{C})$ has a complement L in $\mathcal{C}$ such that $L \cong SL(2,q)$. Clearly L permutes $X = \{X_i \mid 1 \leq i \leq u\}$. Also $L \cap H = \{1\}$ and our theorem follows from [7, Theorem] if q = 2. Thus we assume that q > 2 so that L is simple. If L normalizes each X_i, it follows that G satisfies conclusion (2) of the theorem while if L induces a non-trivial permutation on $\{X_i\}$, it follows that HL has too large a Sylow 2-subgroup. This contradiction completes the proof of

Theorem 1.

We shall now describe the proof of Theorem 3.

PROOF OF THEOREM 3. We assume the hypothesis of Theorem 3 and that $G \neq O(G)C_G(j)$. The methods of [12] for PSp(8,q) carry through [12,(2E]. Thus $\mathcal{C}$ has an elementary abelian subgroup D of order 8 containing j such that $|ccl_G(j) \cap D| = 4$ and such that $N_G(D)$ induces Σ_4 on $ccl_G(j) \cap D$. Then it follows that $N_G(D)$ contains a Sylow 2-subgroup of G. In particular, j is not a central involution of G. Moreover we can then obtain complete knowledge of the G-fusion of the involutions of G. Set $E = D-(ccl_G(j) \cap D)$. Then E is a 4-subgroup of G and we can prove that $O^2(C_G(E)) = C_G(D)$. The method of proof of [12,(2K)] can then be used to determine the structure of the centralizer of an involution of E that is central in G. Also the method of proof of [12,(3C)] can be used to construct a subgroup H of G such that $\mathcal{C} \subseteq H$ and such that H is isomorphic to either Sp(8,q) or PSp(8,q) (using a theorem of Steinberg). Since $\mathcal{C} \subseteq H$, it follows, as in [12,(3A)], that H = G. This forces H = PSp(8,q) and we are done.

Finally we describe the proof of Theorem 4.

PROOF OF THEOREM 4. Assume the hypothesis of Theorem 4 and note that $\mathcal{C}$ = Spin(2n+1,q) has the following two properties:

(1) if t_1,t_2 are involutions of $\mathcal{C}-\langle j\rangle$, then $C_{\mathcal{C}}(t_1) = C_{\mathcal{C}}(t_2)$ if and only if $t_1 \sim t^2$ in $\mathcal{C}$;

(2) if t is an involution of $\mathcal{C}-\langle j\rangle$, then either $C_{\mathcal{C}}(t)$ contains a unique characteristic subgroup J such that $Z(J) = \langle j\rangle$ or $C_{\mathcal{C}}(t)$ = Spin(2n,q) and n is even.

Glauberman's Z*-theorem then implies that there exists an involution $t \in \mathcal{C} -\langle j\rangle$ such that if we set $D = \langle t,j\rangle$, then $C_G(D) = C_{\mathcal{C}}(t) \cong$ Spin(2n,q) and $N_G(D)/C_G(D) = \Sigma_3$. But the outer automorphism group of Spin(2n,q) involves Σ_3 if and only if n = 4 (since $n \geq 4$). Thus n = 4 and there exists a 3-element $y \in N_G(D)$ acting transitively on D#. Also we can determine how y acts on $C_G(D)$ = Spin(8,q). It follows that G has two conjugacy classes of involutions.

Using [11, Theorem], the Thompson transfer lemma and a group order formula of Brauer, we can then determine the structure of the centralizer of involutions not conjugate in G to j. Then we construct a (B,N)-pair in the usual way (cf. [11, sections 2 and 3]) and apply a theorem of Tits [9] to obtain a subgroup H of G such that H is isomorphic to $F_4(q)$. It is then immediate that H = G and the theorem follows.

REFERENCES

[1] M. E. Harris, A characterization of odd order extensions of the finite projective symplectic groups PSp(4,q), Trans. A.M.S., January, 1972.

[2] __________, A characterization of odd order extensions of $G_2(2^n)$ by the centralizer of one central involution, to appear in J. Algebra.

[3] __________, A characterization of odd order extensions of the Steinberg groups $D_4^2(q^3)$, $q = 2^n$ by the centralizer of one central involution, to appear in J. Algebra.

[4] __________, A characterization of odd order extensions of the finite simple projective symplectic group PSp(8,q), to appear.

[5] __________, A characterization of odd order extensions of the finite simple groups $F_4(q)$, q odd, to appear.

[6] E. Shult, On the fusion of an involution in its centralizer, to appear.

[7] G. Thomas, A characterization of the groups $G_2(2^n)$, J. Algebra 13 (1969), 87-118.

[8] __________, A characterization of the Steinberg Groups $D_4{}^2(q^3)$, $q = 2^n$, J. Algebra 14 (1970), 373-385.

[9] J. Tits, Buildings of spherical types and finite (B,N)-pairs, Springer-Verlag, to appear.

[10] W. J. Wong, A characterization of the finite projective symplectic groups $PSp_4(q)$, Trans. Am. Math. Soc. 139 (1969), 1-35.

[11] __________, A characterization of the finite simple groups $PSp_6(q)$, q odd, J. Algebra 12 (1969), 494-524.

[12] __________, Characterization of the finite simple groups $PSp_{2n}(q)$, J. Algebra 14 (1970), 531-551.

REMARK ON SHULT'S GRAPH EXTENSION THEOREM

D. G. Higman

Shult's graph extension theorem [1] fits naturally into the context of <u>regular</u> 2-<u>graphs</u> introduced by G. Higman and studied extensively by D. Taylor [2]. 2-graphs correspond to certain 2-dimensional cocucles and Shult's theorem has an extension to cocycles of arbitrary dimension ≥ 2.

1. COCHAINS

$\Omega[q]$ denotes the set of non-degenerate oriented q-simplices on a finite set Ω, i.e., $\sigma \in \Omega[q]$ is an ordered $(q+1)$-tuple $(\sigma(0),\sigma(1),\ldots,\sigma(q))$ of distinct elements $\sigma(i) \in \Omega$. $C^q(\Omega)$ denotes the set of symmetric functions $f:\Omega[q] \to GF(2)$. The coboundary operator $\delta:C^q \to C^{q+1}$ is defined by

$$(\delta f)(\sigma) = \sum_{i=0}^{q+1} f(\sigma_i) \qquad (f \in C^q, \quad \sigma \in \Omega[q+1])$$

where σ_i is obtained from σ by deleting $\sigma(i)$, so that $\delta^2 = 0$.

For $f \in C^q(\Omega)$, Aut f is the group of permutations π of Ω such that $f(\sigma\pi) = f(\sigma)$ for all $\sigma \in \Omega[q]$, π acting componentwise on σ. Clearly Aut f $\subseteq$ Aut δf.

Choose $x \in \Omega$ and put $\Omega_x = \Omega - \{x\}$. For $f \in C^q(\Omega)$, $q \geq 1$, define $f_x \in C^{q-1}(\Omega_x)$ by $f_x(\mu) = f(x,\mu)$ for $\mu \in \Omega_x[q-1]$. Then

$$(\delta f)(x,\sigma) = (f + \delta f_x)(\sigma)$$

for $\sigma \in \Omega_x[q]$.

The kernel of $\delta:C^q(\Omega) \to C^{q+1}(\Omega)$ is denoted by $Z^q(\Omega)$. If

$f \in Z^q(\Omega)$, $q \geq 1$, then $f = \delta D_x f$, where, for $\mu \in \Omega[q=1]$, $(D_x f)(\mu) = f_x(\mu)$ or 0 according as $\mu \in \Omega_x[q-1]$ or not.

$C^0(\Omega)$ is the set of maps $\Omega \to GF(2)$ and $Z^0(\Omega)$ consists of the two constant maps. $f \in C^1(\Omega)$ corresponds to an ordinary graph on Ω (x and y adjacent if and only if $f(x,y) = 1$) and $f \in Z^1(\Omega)$ means that the corresponding graph is bipartite.

If we refer to a q-simplex σ as <u>coherent</u> for $f \in C^q(\Omega)$ if $f(\sigma) = 1$, then $f \in Z^q(\Omega)$ means precisely that the number of coherent "faces" ρ_i of each (q+1)-simplex ρ is even. Thus the elements of $Z^2(\Omega)$ correspond to 2-<u>graphs</u> [2]: a 2-<u>graph</u> on a set Ω is a set of 3-element subsets of Ω, called <u>coherent</u> <u>triangles</u>, such that the number of coherent triangles amongst the four 3-element subsets of each 4-element subset of Ω is even. Given $f \in Z^2(\Omega)$, the set of $h \in C^1(\Omega)$ such that $f = \delta h$ corresponds to a <u>Seidel</u> <u>class</u> of ordinary graphs (introduced and called <u>switching</u> <u>class</u> by J. J. Seidel). 2-graphs admitting doubly transitive automorphism groups are <u>regular</u> in the sense that any two distinct points lie on the same number of coherent triangles.

2.EXTENSIONS

Choose a new symbol ∞ and put $X = \Omega \cup \{\infty\}$. For $f \in C^q(\Omega)$, $q \geq 1$, define $\tilde{f} \in C^q(X)$ by $\tilde{f}(\infty,\mu) = 0$ for $\mu \in \Omega[q]$ and $\tilde{f}(\rho) = f(\rho)$ for $\rho \in \Omega[q + 1]$, i.e., adjoin ∞ as an "isolated vertex". Put $g = \delta\tilde{f} \in Z^{q+1}(X)$, so that $g(\infty,\sigma) = f(\sigma)$ for $\sigma \in \Omega[q]$ and $g(\rho) = (\delta f)(\rho)$ for $\rho \in \Omega[q + 1]$.

Now choose $x \in \Omega$ and take a permutation π of X such that $\pi(\infty) = x$ and $\pi(x) = \infty$. For $\sigma = (\infty,x,\mu)$, $\mu \in \Omega_x[q - 1]$, $\sigma^\pi = (x,\infty,\mu^\pi)$, so $g(\sigma) = f_x(\mu)$ and $g(\sigma^\pi) = f_x(\mu^\pi)$. Hence $g(\sigma^\pi) = g(\sigma)$ if and only if

(1) $f_x(\mu^\pi) = f_x(\mu)$.

For $\sigma = (x,\rho), \rho \in \Omega_x[q]$, $\sigma^\pi = (\infty,\rho^\pi)$, so $g(\sigma) = (\sigma f)(x,\rho) = (f + \delta f_x)(\rho)$ and $g(\sigma^\pi) = f(\rho^\pi)$. Hence $g(\sigma^\pi) = g(\sigma)$ if and only if

(2) $f(\rho^\pi) = (f + \delta f_x)(\rho)$.

Our remark is that

<u>A permutation</u> π <u>of</u> X <u>which interchanges</u> ∞ <u>and</u> x <u>belongs to</u> Aut g <u>if and only if</u> (1) <u>and</u> (2) <u>hold for all</u> $\mu \in \Omega_x[q-1]$ <u>and all</u> $\rho \in \Omega_x[q]$.

We know that (1) and (2) are necessary. To verify the sufficiency it remains to consider the (q+1)-simplices involving ∞ but not x and those involving neither ∞ nor x.

If $\sigma = (\infty,\rho)$, $\rho \in \Omega_x[q]$, then $\sigma^\pi = (x,\rho^\pi)$, $g(\sigma) = f(\rho)$ and $g(\sigma^\pi) = (\delta f)(x,\rho^\pi) = (f + \delta f_x)(\rho^\pi) = f(\rho^\pi) + (\delta f_x)(\rho)$ (the last step uses (1)). Hence by (2), $g(\sigma) = g(\sigma^\pi)$.

If $\sigma \in \Omega_x[q+1]$,

$$g(\sigma^\pi) = (\delta f)(\alpha^\pi) = \sum_{i=0}^{q+1} f(\sigma^\pi{}_i) = \sum_{i=0}^{q+1} (f + \delta f_x)(\sigma_i)$$

$$= (\delta f)(\sigma) + (\delta^2 f_x)(\sigma) = g(\sigma).$$

The case q = 1 is essentially Shult's graph extension theorem [1]. Using the case q = 2, M_{11} can be constructed as a 3-fold transitive group of degree 12 from PSL(2,11).

REFERENCES

[1] E. Shult, The graph extension theorem. Proc. Amer. Math. Soc. <u>33</u> (1972), 278-284.

[2] Donald Taylor, Some topics in the theory of finite groups. Oxford Thesis 1971.

University of Michigan

SIMPLE GROUPS OF CONJUGATE TYPE RANK + 5

Noboru Ito*

1. NOTATION AND DEFINITION

Let G be a finite group, I(G) the set of indices of centralizers of non-central elements of G in G, and r the number of elements in I(G). r is called the *conjugate type rank of* G. We introduce an ordering in G as follows: let a and b be two elements of I(G). Then a $\geq$ b if and only if a divides b. Let k be the number of maximal elements in I(G). Then G is called k-*headed*. We form a graph C(G) of G as follows: the points of C(G) are the elements of I(G). The (oriented) edge ab of C(G) exists, where a and b are points of C(G), if and only if a > b. We denote the edge ab by

C(G) is called the *conjugate type graph of* G. The centralizer of any non-central element of G in G corresponding to an isolated point of C(G) is called free. G is called *of type* F if all the points of C(G) are isolated.

EXAMPLE. Let LF(2,q) be the linear fractional group over the field of q elements. Then

C(LF(2,4)): 12 15 20
. . .

---

* Partially supported by NSF Grant GP 28420

and

C(LF(2,7)):

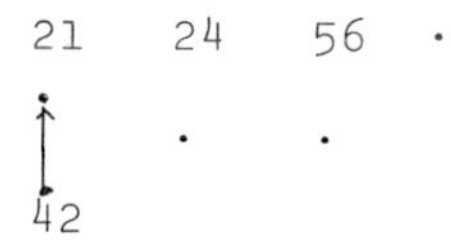

2. PROBLEM

An obvious problem is as follows. Let r be a given positive integer. Then classify all (simple) groups G such that the conjugate type rank of G are equal to r. When r increases, this problem probably will become more difficult with exponential rate. If, however, the shape of C(G) is given and coincident with that of the conjugate type graph of some known simple group, then the problem will become considerably tractable.

3. PRELIMINARIES FOR ATTACKING THE PROBLEM

It is highly desirable to develop a proper method to attack the above problem. Though this seems to be difficult, we have a few leadoffs.

Let g be the order of G and e_i the number of conjugacy classes of elements of G in G whose indices in G are equal to n_i $(i = 1,\dots,r)$, where $I(G) = \{n_1,\dots,n_r\}$. Then we have the equation

(1) $$g = e_0 + e_1 n_1 + \dots + e_r n_r,$$

where e_0 denotes the order of the center of G.

Obviously (1) shows that if G is one-headed then the center of G is non-trivial. So if G is simple (and non-abelian) then G is at least two-headed. But it seems to be probable that there exist no two-headed simple groups. For this see, for instance, Čunihin [4],

Gallagher [6] and Ito [11]. For the present we have to be satisfied with referring to the following lemma, which is more or less known (See Bender [2] and Suzuki [14].

LEMMA. *A subgroup B of a finite group G is called centralizer-closed if for any element B ≠ 1 of B the centralizer of B in G in G is contained in B. Now let G = AB be the product of two subgroups A and B, where the centers of A and B are non-trivial and B is centralizer-closed. Then G is not simple.*

For a proof see Ito [13].

Anyway the two-headed case is the "easiest" to be handled. In general we will rather frequently be led to negotiate with the following situation, which is more elementary and is formulated in the form of a lemma.

LEMMA. *Let H_i be proper subgroups of G and N_i the normalizers of H_i in G (i = 1,...,s). Put $N_i : H_i = m_i$ (i = 1,...,s). If every element of G belongs to some conjugate of H_i for some i, then we have that*

$$\sum_{i=1}^{s} \frac{1}{m_i} > 1. \tag{2}$$

Similar and probably more refined versions of this lemma are maybe abundant in the literature. In this connection, see, for instance, Cunihin [5] and Ito [8].

4. LOCAL CONSIDERATION. A THEOREM OF CAMINA.

First we mention the following lemma:

LEMMA (Ito [8]). *Let H be the centralizer of a non-central element of a group* G *without center. If* H *is free then* H *is a Hall subgroup of* G. *Moreover,* H *is either abelian or of exponent* p *a prime. In particular, if* H *is of even order then* H *is always abelian.*

This lemma obviously shows that a local consideration is sometimes rather helpful in attacking the problem in 2.

Next we would like to notice that the problem in 2 has a loosely inductive character. We start with the following proposition:

PROPOSITION (Ito [8]). *If* G *has the conjugate type rank* 1 *then* G *is nilpotent. Moreover if* I(G) = {n} *then* n *is a power of a prime.*

But this proposition is not so strong and suitable to perform such an inductive function (For this see Ito [9] and [12]). The following theorem of Camina definitely serves for such a purpose.

THEOREM (Camina [3]). *Let* H *be a finite group,* n *an integer and* π *a non-empty set of primes. Let* CsX *denote the centralizer of an element* X *of* H *in* H. *If* H : CsX = n *whenever* X *is a non-central* p'-*element of* H *and* P *is a prime in* π *then either* (i) H *is nilpotent and* n *is a prime power of* (ii) π *consists of only one prime.*

There is a following proposition:

PROPOSITION (Ito [9]). *If* G *has the conjugate type rank* 2 *then* G *is solvable.*

We feel that this proposition should be sharpened for future use along with Camina's line.

5. APPEAL TO THE ESTABLISHED CLASSIFICATION THEOREMS.

Although it is not easy to develop a proper method to attack the problem in §2, it is possible to appeal to the established classification theorems. For instance, the first lemma in §4 shows that circumstantially we can make a direct appeal to the classification theorem of Walter [16], which classifies simple groups with abelian Sylow 2-subgroups. Since the number of this kind of strong "established" classification theorems are ever increasing, we might legitimately hope that we can attack the problem in §2, say, in the case $r \leq 10$. In the case $r \leq 5$ we appeal to Alperin-Brauer-Gorenstein [1], Gorenstein-Walter [7], Suzuki [15], and Wong [17] besides the above-mentioned theorem of Walter.

6. THE CASE $r \leq 4$

We would like to straighten the proofs in Ito [10] and [12].**

THEOREM 1. _A finite group G is a simple group of the conjugate type rank_ 3 _if and only if G is isomorphic with some_ $LF(2,2^m)$, $m \geq 2$.

THEOREM 2. _A finite group G is a simple group of the conjugate type rank_ 4 _if and only if G is isomorphic with some_ $LF(2,q)$, _where_ $q \geq 7$ _is odd_.

REMARK. The case $q = 5$ is an arithmetically degenerate case. Something like this will occur frequently.

** A part of the proof of (2.10) of Ito [12] is not correct. "cyclic" should be replaced by "abelian of rank ≤ 2". It is not difficult to remedy this situation using the second lemma in §3. The part (c) of the proof of (4.8) of that paper is not correct. One way to correct this is shown in (vii)-(xiv) of §7 of this note.

PROOFS. Let G be a simple group of the conjugate type rank $r \leqq 4$. First we show that G cannot be two-headed.

(i) Let n_i be maximal in I(G) (i = 1,2). Let A_i be an element of G such that $G : CsA_i = n_i$, where Cs denotes the centralizer in G (i = 1,2). Then by (1) in §3 $(n_1,n_2) = 1$. In particular $G = CsA_1CsA_2$.

(ii) Both CsA_1 and CsA_2 are not free.

For, by a theorem of Burnside the order of CsA_i is not a prime power. Hence, if CsA_1 is free then CsA_1 is an abelian Hall subgroup of G by the lemma in §4. Therefore CsA_1 is centralizer-closed. Hence G is not simple by the first lemma in §3.

(iii) We may assume that the order of A_i is a prime p_i (i = 1,2). Then $p_1 \neq p_2$, A_i is p_i-central, i.e. belongs to the center of some Sylow p_i-subgroup of G, $n_2 \equiv 0 \pmod{p_1}$ and $n_1 \equiv 0 \pmod{p_2}$.

Since $(n_1,n_2) = 1$, CsA_1 or CsA_2 contains a Sylow p_1-subgroup of G. If CsA_2 contains a Sylow p_1-subgroup of G, then we may assume that A_1 belongs to CsA_2. This implies that CsA_2 contains a normal subgroup of G containing A_1. If $p_1 = p_2$, then we may assume that $A_1A_2 = A_2A_1$.

(iv) We have that either the order of CsA_1 is not divisible by p_2 or the order of CsA_2 is not divisible by p_1.

Assume the contrary. Let $A_2' \neq 1$ be an element of the center of a Sylow p_2-subgroup of CsA_1. We may assume that A_2' belongs to CsA_2. If CsA_2' has the same order as CsA_2, then $G = CsA_1CsA_2'$. Since $A_1A_2' = A_2'A_1$, this implies that G is not simple. If CsA_2' has the same order as CsA_1, then $G = CsA_2'CsA_2$. Since $A_2'A_2 = A_2A_2'$, this implies that G is not simple. If $CsA_1A_2' = CsA_2'$, then A_1 belongs to the center of CsA_2'. Hence $A_1A_2 = A_2A_1$. Then G is not simple.

For the case r = 3 the proof of (iv) ends here.

Now $CsA_1 : CsA_1A'_2$ is prime to p_2. Let P_2 be a Sylow p_2-subgroup of CsA_2. Then we may assume that CsA'_2 contains the center $Z(P_2)$ of P_2. Since $CsA_1A'_2$ contains no conjugates of $Z(P_2)$, we have that $CsA'_2 : CsA_1A'_2$ is divisible by p_2. Hence the order of CsA'_2 does not divide that of CsA_1. Therefore a part of $C(G)$ has the shape

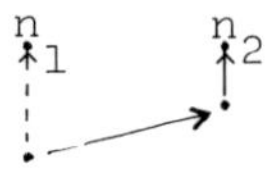

Now by symmetry with respect to n_1 and n_2 we get a contradiction.

(v) We have that the order of CsA_1 is not divisible by p_2 and that the order of CsA_2 is not divisible by p_1.

Assume that the order of CsA_1 is divisible by p_2. Then by (iv) the order of CsA_2 is prime to p_1. Let $A'_2 \neq 1$ be an element of the center of a Sylow p_2-subgroup of CsA_1. Then as in the beginning of the proof of (iv) we obtain that the order of CsA'_2 is not equal to those of CsA_1 and CsA_2, and that $CsA'_2 \neq CsA_1A'_2$. Anyway the order of $CsA_1A'_2$ is divisible by p_1. Further we see that as in the second part of the proof of (iv) the order of CsA'_2 divides that of CsA_2. This is a contradiction.

(vi) CsA_1 and CsA_2 are Hall subgroups of G.

If false, we may assume that there exist a prime q and a Sylow q-subgroup Q of G such that CsA_1 contains Q and $CsA_2 \cap Q \neq 1$. Let $Q \neq 1$ be an element of $CsA_2 \cap Q$. Then the order of CsQ is divisible by p_1p_2. This contradicts (v).

Now we see that $C(G)$ has the shape

(vii) CsA_i is not nilpotent $(i = 1,2)$.

Assume that CsA_1 is nilpotent. Obviously CsA_1 is not abelian. We may assume that the Sylow p_1-subgroup P_1 of CsA_1 is not abelian. Then the Sylow p_1-complement U of CsA_1 is abelian. By a theorem of Burnside $U \neq 1$. Let $X \neq 1$ be an element of CsA_1 of a prime power order. If X belongs to U, then $CsX = CsA_1$. Let X be a p_1-element. By a theorem of Wielandt CsX is nilpotent. Hence CsX is contained in $CsU = CsA_1$. Thus CsA_1 is centralizer closed. Hence G is not simple.

The following argument is due to Camina [3].

(viii) Let $n_1 > n_3$ and $n_2 > n_4$. In Camina's theorem let $H = CsA_1$ π the set of prime divisors of the order of the center of CsA_1. Then we see that n_3/n_1 is a power of p_1. Similarly n_4/n_2 is a power of p_2. By a theorem of Burnside let $q_1 \neq p_1$ and $q_2 \neq p_2$ be prime divisors of orders of CsA_1 and CsA_2 respectively. Then for every element X of $G, G : CsX$ is prime to either q_1 or q_2. Hence G is not simple (Ito [8]; c.f. the second lemma in §3).

7. THE CASE $r \leqq 4$ CONTINUED

Now we are in the three-headed case. If $r = 3$ then G is of type F. So a Sylow 2-subgroup of G is abelian. By Walter's theorem [16] G is isomorphic with some $LF(2,2^m)$, $m \geqq 2$. Hence from now on we assume that $r = 4$.

Let n_i be the maximal in I(G) $(i = 1,2,3)$. Let A_i be an element of G such that $G : CsA_i = n_i$ $(i = 1,2,3)$. Our first goal is to show that at least one of CsA_i is free.

Assume the contrary. Then let X_i be an element of G such that CsX_i is properly contained in $CsA_i (i = 1,2,3)$. Then $G : CsX_i = n_4$. Obviously CsA_i is not nilpotent. Hence by Camina's theorem we obtain that $CsA_i : CsX_i = p_i^{a_i}$, where p_i is a prime, and that the center of CsA_i is a p_i-group ($i = 1,2,3$). By the choice of A_i, the p_i's are distinct.

(i) The order of CsA_i is divisible only by p_1, p_2 and p_3 ($i = 1,2,3$).

Let q be a prime divisor of the order of CsA_1 distinct from p_i's. We may assume that CsA_1 contains a Sylow q-subgroup Q of G. Let $Q \neq 1$ be an element of the center of Q. Then we have that CsA_1Q and its conjugates exhaust G. Hence G is not simple.

(ii) If $G : CsX = G : CsX_1$ then CsX is abelian.

This is obvious, since $CsA_i : CsX_i$ is a power of p_i.

(iii) We may choose $X = X_1$ and A_i ($i = 1,2,3$) so that CsX is contained in $\bigcap_{i=1}^{3} CsA_i$.

(iv) CsA_i is solvable ($i = 1,2,3$). For at least one i the Fitting subgroup of CsA_i is a p_i-group.

(v) For at least two i's the Fitting subgroup of CsA_i is a p_i-group.

(vi) If the Fitting subgroup of CsA_3 is not a p_3-group, then $p_3 < p_i$ ($i = 1,2$).

We may assume that $p_1 > p_2 > p_3$. Then it is easy to show that a Sylow p_1-subgroup of G is normal in G.

For proofs of (iii) - conclusion see Ito [12].

Now we may assume that CsA_3 is free. Our next goal is to show that at least one of CsA_i ($i = 1,2$) is also free.

Assume the contrary. Then let X_i be an element of G such that CsX_i is properly contained in CsA_i ($i = 1,2$). Then $G : CsX_i = n_4$. Obviously CsA_i is not nilpotent ($i = 1,2$). Hence by Camina's theorem we obtain that $CsA_i : CsX_i = p_i^{a_i}$, where p_i is a prime, and that the center of CsA_i is a p_i-group ($i = 1,2$). By the choice of

A_i, p_i's are distinct.

(vii) The order of CsA_i is divisible only by p_1 and p_2.

Let q be a prime divisor of the order of CsA_1 distinct from p_i (i = 1, 2). We may assume that CsA_1 contains a Sylow q-subgroup Q of G. Let $Q \neq 1$ be an element of the center of Q. Then we have that CsA_1Q contains Q and that $G : CsA_1Q = G : CsX_1$. This shows that G is not simple (Ito [8]).

(viii) p_1 or p_2 is equal to 2.

As assume that $p_2 = 2$. Then by Walter's theorem [16] a Sylow 2-subgroup P_2 of CsA_2 is not abelian, and, in particular, is of exponent $\geqq 4$.

(ix) There exists a 2-element X such that $G : CsX = G : CsX_1$. CsX is the direct product of the Sylow 2-subgroup P^*_2 and the abelian Sylow p_1-subgroup P^*_1.

(x) We may choose A_i so that $CsA_1 \cap CsA_2 = CsX$.

(xi) The Fitting subgroup $F(CsA_2)$ of CsA_2 is not a 2-group.

For the proofs of (viii) - (xi) see Ito [12].

Now $F(CsA_2) = CsX = P^*_1 \times P^*_2$. Since P_2CsX/P^*_2 is a Frobenius group with CsX/P^*_2 the kernel, P_2/P^*_2 is cyclic or generalized quaternion. Let A'_2 be an element of $P_2 - P^*_2$. If $G : CsA'_2 = G : CsX$, then A_2 commutes with a p_1-element not belonging to P^*_1. Hence $G : CsA'_2 = G : CsA_2$. If CsA'_2 contains a 2-element X' of CsA_2 such that CsX' = CsX, then A'_2 belongs to CsX. This is a contradiction. Hence CsA'_2 does not contain such an element. If $\langle A'_2 \rangle \cap P^*_2 \neq 1$, then CsA'_2 contains P^*_1. This is a contradiction. Hence $\langle A'_2 \rangle \cap P^*_2 = 1$.

(xii) The center $Z(P_2)$ of P_2 is elementary abelian.

First show that the center $Z(CsA_2)$ is elementary abelian. Otherwise, we may assume that A_2 is an element of order 4. Let A'_2

be an involution of $P_2 - P_2^*$. Then $CsA_2'A_2 = CsA_2^2 = CsA_2$. This shows that A_2' belongs to $Z(CsA_2)$, and hence, to CsX. This is a contradiction. Now assume that $Z(P_2)$ is not elementary abelian. Let A_2' be an element of $Z(P_2)$ of order 4. Then A_2' does not belong to P_2^* by the first argument. But $A_2'X = XA_2'$. This is a contradiction.

(xiii) $P_2 : P_2^* = 2$.

By (xii) and by the argument before that this is obvious.

(xiv) $CsA_2' \cap P_2^* = \langle A_2 \rangle$.

Let A_2' be an element of $Z(P_2^G)$, where G is an element of G. Then A_2^G belongs to $Z(P_2^G)$. $CsA_2^G : CsX^G = 2$ and $CsX^G = P_2^{*G}$. Then A_2' belongs to P_2^{*G}. Hence $CsA_2' = CsA_2^G$. Now assume that $CsA_2' \cap P_2^*$ contains $\langle A_2 \rangle$ properly. Then CsX^G contains an element $A_2'' \neq 1$ of P_2^*. Then CsA_2'' contains P_1^* and P_1^{*G}. The first argument shows that the Fitting subgroup $F(CsA_2'')$ of CsA_2'' is not a 2-group. Hence $P_1^* = P_1^{*G}$. This is a contradiction.

Now by a lemma of Suzuki P_2 is dihedral or quasidihedral. Hence by a theorem of Gorenstein-Walter [7] or by a theorem of Alperin-Brauer-Gorenstein [1] we get a contradiction.

Finally we may assume that CsA_2 and CsA_3 are free. Our goal is to prove that G is isomorphic with some $LF(2,q)$, where $q \geqq 7$ is odd.

Since CsA_1 cannot be free, let X be an element of G such that $CsA_1 : CsX = n_4$.

(xv) CsA_1 is a Hall subgroup of G.

This is obvious.

(xvi) CsA_1 is of even order.

(xvii) We may assume that CsA_1 is not nilpotent.

For these two see Ito [12].

Now by Camina's theorem $CsA_1 : CsX = p^a$, where p is a prime, and the center of CsA_1 is a p-group.

(xviii) We may assume that $p = 2$.

Otherwise let J be an involution in CsA_1. Then $CsJ = CsA_1J$ is nilpotent. Then by the proof of (xvii) we may assume that $p = 2$.

By (xvii) CsA_1 is not a 2-group. Anyway, as before, P_1 is not abelian and, in particular, is of exponent $\geqq 4$.

(xix) There exists a 2-element Y such that $G : CsY = G : CsX$.

(xx) The Fitting subgroup $F(CsA_1)$ is not a 2-group.

For these two see (ix) and (xi).

Now the theorems of Suzuki [15] and Wong [17] enable us to show that $F(CsA_1)$ is a 2-group, unless G is isomorphic with some LF(2,q), $q \geqq 7$, odd. This is the argument of Ito [12], (2.10), which can be simplified because of Camina's theorem.

8. THE CASE r = 5

Elaborating the arguments in §7 and §6 we can prove the following theorem:

THEOREM. (i) _There exists no two-headed simple group of the conjugate type rank_ 5.

(ii) _A finite group_ G _is a four-headed simple group of the conjugate type rank_ 5 _if and only if_ G _is isomorphic with some Suzuki group or_ LF(3,4).

REFERENCES

[1] J. L. Alperin, R. Brauer, D. Gorenstein, Finite groups with quasidihedral and wreathed Sylow 2-subgroups, Trans. Amer. Math. Soc. 151 (1970), 1-260.

[2] H. Bender, Transitive Gruppen gerader Ordnung, in denen jede Involutioned genau einen Punkt festläst. J. Algebra 17 (1971), 527-556.

[3] A. R. Camina, Conjugacy Classes of finite groups and some theorems of N. Ito. To appear in J. London Math. Soc.

[4] S. A. Čunihin, Simplicité de groupe fini et les ordres de ses classes d'éléments conjugués, Comptes Rendu Acad. Sci. Paris 191 (1950), 397-399.

[5] __________, Sur les probleme des deux classes d'un group fini, Compte Rendu Acad. Sci. Paris 198 (1934), 531-532.

[6] P. X. Gallagher, The conjugacy classes in a finite simple group, J. für reine und angewandt Math. 239/240 (1970), 363-365.

[7] D. Gorenstein and J. H. Walter, On finite groups with dihedral Sylow 2-subgroups, Illinois J. Math. 6 (1962), 335-357.

[8] N. Ito, On finite groups with given conjugate types I, Nagoya Math. J. 6 (1953), 17-28.

[9] __________, On finite groups with given conjugate types II, Osaka J. Math. 7 (1970), 231-251.

[10] __________, On finite groups with given conjugate types III, Math. Zeit. 117 (1970), 267-271.

[11] __________, On factorizable groups, Proc. Symp. Pure Math. 21 (1971), 77-83.

[12] __________, Simple groups of conjugate type rank 4, J. Algebra 20 (1972).

[13] __________, A theorem on factorizable groups, to appear in Acta sci. math. Szeged.

[14] M. Suzuki, Two characteristic properties of (ZT)-groups, Osaka Math. J. 15 (1963), 143-150.

[15] __________, Finite groups in which the centralizer of any element of order 2 is 2-closed, Ann. Math. 82 (1965), 191-212.

[16] J. H. Walter, The characterization of finite groups with abelian Sylow 2-subgroups, Ann. Math. 89 (1969), 405-514.

[17] W. Wong, On finite groups whose 2-Sylow subgroups have cyclic subgroups of index 2, J. Australian Math. Soc. 4 (1964), 90-112.

University of Illinois at Chicago Circle

A CLASS OF SIMPLE GROUPS OF CHARACTERISTIC 2

Zvonimir Janko

We shall state at first the following:

PROBLEM A. Determine the structure of any non-abelian finite simple group G which possesses an involution z lying in the center of an S_2-subgroup of G such that the centralizer H of z in G has the following properties:

(i) The subgroup $E = O_2(H)$ is of class at most 2.

(ii) The group H possesses a normal subgroup H_0 of odd index such that H_0/E is isomorphic to $L_2(2^n)$, $n \geq 2$.

(iii) We have $C_H(E)$ _ E.

It seems that the following simple groups are the only known finite simple groups satisfying the assumptions of Problem A: $L_4(2^n)$, $U_4(2^n)$, $Sp_4(2^n)$, $G_2(2^n)$, $D_4^2(2^n)$, $n \geq 2$, J_2 and J_3.

To solve the above problem, we start with a result of Shult which implies that the weak closure of $\langle z \rangle$ in H is not abelian. This gives an initial fusion of involutions in G.

Even a special case of the above problem seems to be quite difficult. This is seen from the following result.

THEOREM A. *Let G be a nonabelian finite simple group satisfying the assumptions of Problem A. Assume in addition that E is an extra-special group. Then G is either isomorphic to one of the following groups:* J_2, J_3, $D_4^2(8)$ *or G is a simple group with the following properties: The subgroup E is the central product of n + 1 quaternion groups where n is odd and* $n \geq 5$. *We have* H_0 *is equal to*

H. *If* P *is a cyclic subgroup of order* $2^n - 1$ *of* H, *then* $C_E(P) = C_E(x)$ *is a dihedral group of order* 8, *where* x *is any nonidentity element in* P. *If* Q *is a cyclic subgroup of order* $(2^n + 1)/3$ *of* H, *then* $C_E(Q) = C_E(y)$ *is a quaternion group, where* y *is any nonidentity element in* Q. *The involution* z *is fused in* G *to an involution* z' *in* E - <z> *but* z *is not fused to any involution in* H - E. *The group* G *has precisely two conjugate classes of involutions with the representatives* z *and* $z_1 \in E - \langle z \rangle$. *The group* H *splits over* E. *The structures of* H *and* $H_1 = C_G(z_1)$ *are uniquely determined. The group* H_1 *is solvable of order* $2^{3n+1}.3$, *where* $O(H_1) = 1$ *and* $|O_2(H_1)| = 2^{3n}$. *Every* 2-*local subgroup of* G *has cyclic Sylow* p-*subgroups for all odd primes* p. *Finally, the order of* G *is*:

$$|G| = 2^{3n+3}.3.(2^n+1)(2^n-1)(2^{n+1}+2^{1/2(n+1)}+1)(2^{n+1}-2^{1/2(n+1)}+1),$$

which implies $n = 3m_0$ (*since* $9 \mid (2^n + 1)$).

The proof of Theorem A is obtained in the following way. We show at first that the involution z is fused in G to an involution z' in E - <z>. Let E be of order 2^{2m+1}. If $m \leq 4$, then we show easily that G must be isomorphic to J_2 or J_3 or $D_4^2(8)$. Hence we may assume that $m \geq 5$. The above fusion result has many consequences. First of all, we get $n \geq m - 1$ and the first part of the theorem follows quickly. The fusion problem for involutions in G is then solved in the standard way. In particular, we get that the group G has precisely two classes of involutions with the representatives z and $z_1 \in E - \langle z \rangle$. In the determination of the structure of $H_1 = C_G(z_1)$ we also use the above mentioned result of Shult. The order of G is then determined by the order formula of J. G. Thompson.

We get $n = 3m_0 = m - 1$, m_0 is odd, and the group G does not exist for $m_0 = 3$. What about $m_0 > 3$?

The Ohio State University

2-GROUPS WHICH CONTAIN EXACTLY THREE INVOLUTIONS

Marc W. Konvisser*

INTRODUCTION

A well known result in the theory of finite p-groups is that a p-group which has only one subgroup of order p is either cyclic or generalized quaternion [4, §§104-5]. The problem of classifying p-groups which contain exactly one elementary subgroup of order p^2 has been done (at least implicitly) for all cases except p = 2. The results for $p \neq 2$ are outlined below, while the main body of this paper concerns the classification 2-groups which contain exactly one elementary subgroup of order 2^2 (i.e., 2-groups which contain exactly 3 involutions).

If p > 3, then the class of all p-groups which contain exactly one elementary subgroup of order p^2 is the class of all non-cyclic metacyclic p-groups. This follows directly from [2, Theorem 3.2].

If G is a 3-group which contains exactly one elementary subgroup of order 3^2, then either G is a non-cyclic metacyclic 3-group or a 3-group of maximal class. The fact that G may have maximal class follows from [2, Theorem 4.1]; the 3-groups of maximal class are classified in [3, p. 88].

So the main problem lies in classifying those 2-groups which contain exactly one elementary subgroup of order 4.

In what follows we will show that a 2-group which contains exactly 3 involutions contains a metacyclic subgroup of index at

---

*This work was supported in part by a faculty research grant from Illinois State University

most 4 and that this subgroup is normal in most cases.

Our method is to classify the metacyclic 2-groups which contain exactly 3 involutions first, and then to show that if G is a non-metacyclic 2-group which contains exactly 3 involutions then G has a metacyclic subgroup of small index. We then analyze the possible extensions of this metacyclic subgroup to get our final result. In §4 we give examples to show that our results are "best possible."

NOTATION AND DEFINITIONS

1. Definition: G is a metacyclic group if and only if G is a cyclic extension of a cyclic group, cf. [2].
2. Definition: An element of a group of order 2 is called an involution. In addition to the standard group theoretical notation [cf.5] we will also use:
3. $A \leq B$, A is a subgroup of B.
4. $\Omega_i(G)$, the subgroup of G generated by all elements of G of order dividing p^i.
5. $\mho^i(G)$, the subgroup of G generated by all p^i-th powers of elements of G.
6. $A_G(W)$, the subgroup of the automorphism group of W isomorphic to $G/C_G(W)$.
7. 2-Aut(W), a Sylow 2-subgroup of the automorphism group of W.
8. $a^g = g^{-1}ag$
9. $[a,g] = a^{-1}g^{-1}ag$.

1. METACYCLIC GROUPS

1.1 THEOREM. Let G be a metacyclic 2-group. Then

1. G contains exactly one involution if and only if G is either cyclic or generalized quaternion.
2. G contains more than 3 involutions if and only if G is either dihedral (of order greater than 4) or semi-dihedral.
3. All other metacyclic 2-groups contain exactly 3 involutions.

2. IF G HAS NO NORMAL $Z_4 \times Z_4$

We divide the non-metacyclic 2-groups which contain exactly 3 involutions into two classes--those which contain a normal subgroup isomorphic to $Z_4 \times Z_4$ and those which do not.

2.1 THEOREM. Let G be a non-metacyclic finite 2-group containing exactly 3 involutions which has no normal subgroup isomorphic to $Z_4 \times Z_4$. Then G contains a normal subgroup $Y \simeq Z_4 \times Z_2$, $C_G(Y) \simeq Z_{2^n} \times Z_2$ where $n \geq 2$; and if $n \geq 3$, then $|G/C_G(Y)|$ divides 4, and if $n = 2$ then $G/C_G(Y)$ is isomorphic to a subgroup of the dihedral group of order 8.

3. IF G CONTAINS A NORMAL $Z_4 \times Z_4$

ASSUMPTION FOR §3. G will always denote a non-metacyclic 2-group containing exactly 3 involutions which has a normal subgroup $W \simeq Z_4 \times Z_4$.

3.1 LEMMA. Let A be a 2-group containing exactly 3 involutions such that $\Omega_2(A) \simeq Z_4 \times Z_4$. Then A is metacyclic.

3.2 THEOREM. *Let G be a 2-group containing exactly 3 involutions and a normal subgroup* $W \cong Z_4 \times Z_4$. *If* $\Omega_1(W) \leq Z(G)$, *then G contains a normal metacyclic subgroup A of index at most 4.*

Next, we consider the group G for which $\Omega_1(W) \not\leq Z(G)$.

3.3 LEMMA. *If* $\Omega_1(W) \not\leq Z(G)$, *then, with appropriate choice of generators u and y of W, one of the following automorphisms is contained in* $A_G(W)$.

$$\lambda(i,j):\quad u \to uy$$
$$y \to u^{2i}y^j \quad \text{where } i \in Z_2 \text{ and } j \equiv 1 \pmod 2.$$

3.4 THEOREM. *If* $|A_G(W)| = 2^4$, *then* $W = C_G(W)$ *and* $A_G(W) = \langle\lambda(0,1), \lambda(1,-1)\rangle$.

3.5 LEMMA. *If* $|A_G(W)| = 8$, *then G contains a metacyclic subgroup M of index dividing 4, and M is normal in G except in the case in which* $A_G(W) = \langle\lambda(1,-1), \lambda(0,-1)\rangle$ *or* $A_G(W) = \langle\lambda(1,-1), \lambda^2(1,1)\rangle$.

4. EXAMPLES

In this section we give examples of some families of 2-groups which contain exactly 3 involutions. Although a complete classification is not given, some of the more complicated cases are listed.

4.1 We construct a group G in which $G/Y \cong \mathrm{Aut}(Y)$, G has no normal $Z_4 \times Z_4$ and G contains exactly 3 involutions, cf., 2.1.

Let H be the non-split extension of

$$Y = \langle y, u_0 \mid y^4 = u_0^2 = [y,u_0] = 1\rangle$$

by an element g given by

$$H = \langle g, y, u_0 \mid g^8 = u_0^2 = [y,u_0] = 1,\ y^2 = g^4,\ y^g = yu_0,\ u_0^g = u_0y^2\rangle$$

Then $\Omega_1(H) = \Omega_1(<g^2,Y>) = <y^2,u_0>$ since $(g^2y^iu^j)^2 = g^4 = y^2$.

Now let G be the non-split extension of H by a given by

$$G = <a,H|a^2 = y,\ u_0^a = u_oy^2,\ g^a = g^{-1}>.$$

One can easily check that the group G exists. Now we note that

$$(ay^iu_0^j)^2 = a^2y^{2i+2j} = y^{1+2(i+j)}$$

$$(ag)^2 = a^2g^{-1}g = a^2$$

$$(agy^iu_0^j)^2 = (ag)^2y^{2i}u_0^i$$

So every element of the cosets aY and agY has order 8 and has its fourth power equal to y^2; since $a^g = ag^2$ and $(ag)^8 = (ag^{-1})$, we see that this is also true for all elements of G-H.

Thus $\Omega_1(G) = <y^2,u_0>$ and G has no normal $Z_4 \times Z_4$.

4.2. The simplest example of a family of groups which contains a normal $Z_4 \times Z_4$ is the direct product of two generalized quaternion groups. This group contains a maximal metacyclic (abelian) subgroup of index 4, namely the direct product of the cyclic maximal subgroups of each factor.

4.3. We construct a family of groups for which $A_G(W) = <\lambda(0,-1), \lambda(1,-1)>$ (cf. 3.5).

We begin by constructing the non-split extension K of the abelian group A where

$$A = <a,b|a^{2n} = n^{2n-1} = [a,b] = 1> \text{ for } n \geq 3$$

by an element k such that

$$a^k = ab,\ b^k = a^{-2}b^{-1},\ k^4 = a^{2^{n-1}}.$$

We then extend K by an element g such that

$$a^g = a, b^g = a^{-2}b^{-1}, \; k^g = k^{-1}a^{1+2^{n-1}}, \; g^2 = a.$$

The group $G = \langle g,K\rangle$ with the defining relations given above represents a group that contains a metacyclic subgroup of index 4, namely $\langle g,A\rangle = \langle g,bg^2\rangle$, but no normal metacyclic subgroup of index 4.

4.4. We now construct a group G for which $A_G(W) = \langle\lambda(0,1), \lambda(1,-1)\rangle$ (cf., 3.4).

We begin by constructing an extension K of W where

$$K = \langle k,u \mid k^8 = u^4 = 1, \; u^k = u^{-1}y, \; y^k = u^2y, \; [u,y] = 1 \text{ and } y^2 = k^4\rangle, \quad \text{(cf. [6, 2.2])}.$$

We extend K by an element s which induces $\lambda^2(0,1)$ on W to form the group S defined by

$$S = \langle s,K \mid s^2 = y, \; u^s = uy^2, \; k^s = kuy\rangle.$$

Using [6, 2.5.1] we see that such a group can be constructed.

Finally we construct G as an extension of S by an element t such that $t^2 = s$ where G is defined by $G = \langle t,s \mid t^2 = s, \; k^t = k^{-1}suy, \; u^t = uy\rangle$. Using calculations identical to those in [6,2.6.1] we see that such a group G exists.

REFERENCES

[1] J. Alperin, Centralizers of abelian normal subgroups, J. Algebra 1 (1964),110-113.

[2] N. Blackburn, Generalizations of certain elementary theorems on p-groups, Proc. Lond. Math. Soc. (3) 11 (1961), 1-22.

[3] __________, On a special class of p-groups, Acta Math. 100 (1958), 45-92.

[4] W. Burnside, Theory of groups of finite order, second ed., Cambridge 1911 (Reprinted 1955, Dover Publications, New York).

[5] B. Huppert, Endliche Gruppen I, Springer-Verlag (Berlin, 1967).

[6] M. Konvisser, 2-groups of normal rank 2 for which the Frattini subgroups has rank 3, Trans. Amer. Math. Soc. (to appear)

Wayne State University

ON SOLVING THE DEGREE EQUATIONS IN π-GROUPS

Jeffrey S. Leon

Let π be a fixed, finite set of primes. A π-number is a positive integer each of whose prime factors is in π. A π-group is a finite group G such that $|G|$ is a π-number.

Given a π-group G, the theory of blocks yields diophantine equations in which the unknowns are degrees of irreducible characters and consequently π-numbers. For example, if $p \in \pi$ and a Sylow p-subgroup P of G is cyclic of order p^c, we obtain the equations

$$1 + \varepsilon_2 z_2 + \ldots + \varepsilon_e z_e + \varepsilon_0 z_0 = 0,$$

$$\varepsilon_i = \pm 1, \quad z_i \equiv \varepsilon_i \pmod{p^c} \text{ for } i = 2,3,\ldots,e,$$

(1) $$z_0 \equiv -e\varepsilon_0 \pmod{p^c}, \; z_i \text{ is a } \pi^*\text{-number, where}$$

$$\pi^* = \pi - \{p\}.$$

Here $e = |N_G(P):C_G(P)|$, $1, z_2, \ldots, z_e$ are the degrees of the nonexceptional characters in the principal p-block, and z_0 is the common degree of the exceptional characters. Even in the case of a noncyclic Sylow p-subgroup, the theory of exceptional characters may yield equations of the same general type as (1).

It turns out that in some (if not all) cases, the equations (1) have only a finite number of solutions. A method for solving equations of this type can be useful in classifying -groups. We shall outline such a procedure here. The method is elementary in that it depends only on the theory of congruences. Because of the large number of calculations involved, use of a computer is necessary in all but the simplest cases. However, the algorithms are easily programmed in a user-oriented language such as FORTRAN.

For simplicity, we shall consider diophantine equations in which the unknowns are π-numbers, where $\pi = \{p,q\}$ has cardinality two. (This is always the case if $|G| = p^a q^b r^c$ and if the Sylow r-subgroups of G are either cyclic or else elementary abelian of order r^2). Generalization to higher values of $|\pi|$ is straightforward. Explicitly, we shall consider the equation:

(*) $$a_1 p^{x_1} a^{y_1} + a_2 P^{x_2} q^{y_2} + \ldots + a_n P^{x_n} q^{y_n} = b,$$
(other conditions).

Here the x_i and y_i are the unknowns. The exact nature of the "other conditions" is unimportant here, except that they must insure that the following hold:

(**) If $x_1, \ldots, x_n, y_1, \ldots, y_n$ is any solution to (*) and if T is any nonempty subset of $\{1,2,\ldots,n\}$, then $\sum_{i \in T} a_i p^{x_i} q^{y_i} \neq 0$. In particular, $b \neq 0$.

Note that in the equations (1), the congruences for the z_i insure that (**) holds.

A solution of (*) will be denoted by $(x_1, \ldots, x_n, y_1, \ldots, y_n)$ and abbreviated by $(\overline{x},\overline{y})$. $\overline{x} \equiv \overline{x'}$ (mod u) and $\overline{y} \equiv \overline{y'}$ (mod v) will mean the following: $x_i \equiv x'_i$ (mod u) and $y_i \equiv y'_i$ (mod v) for all i.

Consider a particular case of (*), the degree equation for a $\{2,3,5\}$-group with cyclic Sylow 5-group R and $|N(R):C(R)| = 4$:

$$\varepsilon_1 2^{x_1} 3^{y_1} + \varepsilon_2 2^{x_2} 3^{y_2} + \varepsilon_3 2^{x_3} 3^{y_3} + \varepsilon_4 2^{x_4} 3^{y_4} = -1, \tag{2}$$

$$\varepsilon_i = \pm 1, \quad 2^{x_i} 3^{y_i} \equiv \varepsilon_i \pmod 5).$$

It is straightforward (by computer) to find all solutions to the above equations in which each term $2^{x_i}3^{y_i}$ is small, say less than 10^5. We obtain 21 solutions, listed in Table 1.† Placing a higher bound of the terms $2^{x_i}3^{y_i}$, say 10^7 or 10^9, we still obtain only the 21 solutions in Table 1. We are led to conjecture that the list of solutions in Table 1 is the complete list of solutions of the equations. How can we prove this?

More generally, given a set S of solutions to (*), how can one prove that S is the complete set of solutions of (*)? The theorem which follows says that it will suffice to prove that any solution is "congruent" to a solution in S in an appropriate modulus.

DEFINITION. Let $S = \{ (\overline{x}_i, \overline{y}_i) \mid i = 1,2,\ldots,m\}$ be a set of solutions of (*). Here $(\overline{x}_i, \overline{y}_i) = (x_{i1}, x_{i2}, \ldots, x_{in}, y_{i1}, \ldots, y_{in})$. $\nu p(k)$ denotes the largest integer h such that p^h divides k.

$$\nu_p(S) = 1 + \max \{ \nu_p(\sum_{j \in T} a_j p^{x_{ij}} q^{y_{ij}}) \mid i = 1,2,\ldots,m;$$
$$T \subseteq \{1,2,\ldots,n\},\ T \neq \phi\}.$$

$\nu_q(S)$ = (same as above except ν_q in place of ν_p),

$\tau_p(S)$ = exponent of q modulo $p^{\nu_p(S)}$) = smallest integer k such that $q^k \equiv 1 \pmod{p^{\nu_p(S)}}$,

$\mu_p(S)$ = Least multiple of $\tau_p(S)$ which is $\geq \nu_q(S)$.

---

† More precisely, there are 24·21 solutions, corresponding to all possible arrangements of the terms in each equation. In practice, extensive use is made of symmetry assumptions.

Note that (**) is necessary to insure that $\nu_p(S)$ and $\nu_q(S)$ are finite. For the set S of solutions in Table 1 (with p = 3, q = 2), $\nu_3(S) = 8$, $\nu_2(S) = 15$, $\tau_3(S) = 2\cdot 3^7$, and $\mu_3(S) = 2\cdot 3^7$.

THEOREM. <u>Let</u> $S = \{ (\overline{x}_i,\overline{y}_i) \mid i = 1,2,\ldots,m$ <u>be a set of solutions to</u> (*) <u>with the following property</u>: <u>If</u> $(\overline{x},\overline{y})$ <u>is any solution to</u> (*), <u>then there is an integer</u> i <u>such that</u>

<u>i</u> $\overline{x} \equiv \overline{x}_i \pmod{w}$ <u>for some</u> $w \geq \nu_p(S)$

<u>and</u> <u>ii</u> $\overline{y} \equiv \overline{y}_i \pmod{\mu_p(S)}$.

<u>Then</u> $(\overline{x},\overline{y}) = (\overline{x}_i,\overline{y}_i)$ <u>and</u> S <u>is the complete set of solutions of</u> (*).

PROOF. Write $\overline{x} = (x_1,\ldots,x_n)$, $\overline{y} = (y_1,\ldots,y_n)$, $\overline{x}_i = (x_{i1},\ldots,x_{in})$, and $\overline{y}_i = (y_{i1},\ldots,y_{in})$. We are given that $x_k \equiv x_{ik} \pmod{w}$ and $y_k \equiv y_{ik} \pmod{\mu_p(S)}$ for all k.

Let $N = \{1,2,\ldots,n\}$ and let $T = \{j\varepsilon N \mid x_j = x_{ij}\}$. The definition of $\nu_p(S)$ implies that $x_{ij} \leq \nu_p(S)$ for all j; then condition <u>i</u> implies that $p^{x_j} \equiv 0 \pmod{p^{\nu_p(S)}}$ if $j \notin T$. By the definition of μ_p and condition <u>ii</u>, $q^{y_j} \equiv q^{y_{ij}} \pmod{p^{\nu_p(S)}}$ for all j.

$$\begin{aligned} 0 &= \sum_{j=1}^{n} a_j p^{x_j} q^{y_j} \\ &\equiv \sum_{j\varepsilon T} a_j p^{x_j} q^{y_j} \pmod{p^{\nu_p(S)}} \\ &\equiv \sum_{j\varepsilon T} a_j p^{x_{ij}} q^{y_{ij}} \pmod{p^{\nu_p(S)}} \\ &= \sum_{j=1}^{n} a_j p^{x_{ij}} q^{y_{ij}} - \sum_{j\varepsilon N-T} a_j p^{x_{ij}} q^{y_{ij}} \\ &= -\sum_{j\varepsilon N-T} a_j p^{x_{ij}} q^{y_{ij}} . \end{aligned}$$

Thus $p^{\nu_p(S)}$ divides $\sum_{j\varepsilon N-T} a_j p^{x_j} q^{y_j}$, which contradicts the definition of $\nu_p(S)$ unless $N-T = \phi$. Thus $x_j = x_{ij}$ for all j.

Now let $U = \{ j\varepsilon N \mid y_j = y_{ij} \}$. $y_{ij} \leq \nu_q(S) \leq \mu_p(S)$ for all j, and by condition ii, $q^{y_j} \equiv 0 \pmod{q^{\nu_q(S)}}$ if $j \notin U$. Proceeding as before with p and q interchanged, we obtain $N-U = \phi$ and $y_j = y_{ij}$ for all j, completing the proof.

Thus the task of proving S the complete set of solutions of (*) reduces to showing that i and ii hold. Again, the method requires only the theory of congruences. We require sequences $\{u_i\}_{i=0}^{\ell}$, $\{v_i\}_{i=0}^{\ell}$, and $\{P_i\}_{i=1}^{\ell}$ as follows:

a) $\{u_i\}$ and $\{v_i\}$ are sequences of positive integers such that $u_0 = v_0 = 1$, u_i divides u_{i+1} and v_i divides v_{i+1} for $i = 0,1,\ldots,\ell-1$, $\nu_p(S)$ divides u_ℓ, and $\mu_p(S)$ divides v_ℓ.

b) Each P_i is a set of numbers each of which is a prime power (in practice, usually either a prime or a power of p or q).

c) If $r \varepsilon P_i$ and r is not a power of p or q, then $p^{u_i} \equiv q^{v_i} \equiv 1 \pmod r$. If $p^j \varepsilon P_i$, then $j \leq u_i$ and $q^{v_i} \equiv 1 \pmod{p^j}$. If $q^k \varepsilon P_i$, then $k \leq v_i$ and $p^{u_i} \equiv 1 \pmod{q^k}$.

Let $S_0 = (\overline{0},\overline{0})$, where $\overline{0}$ is the n-vector of zeroes. $S_1, S_2,\ldots, S_\ell$ are constructed inductively as follows. Assume that $S_k = \{ (s_{j1},\ldots,s_{jn},t_{j1},\ldots,t_{jn}) \mid j = 1,2,\ldots,m_k \}$. Let $c = u_{k+1}/u_k$ and $d = v_{k+1}/v_k$. Define the set $S_{k+1}{}^*$ by $S_{k+1}{}^* = \{ (s_{j1}+h_1u_k,\ldots, s_{jn}+h_nu_k, t_{j1}+g_1v_k,\ldots, t_{jn}+g_nv_k \mid j = 1,2,\ldots,m_k;\ h_1,h_2,\ldots,h_n = 0,1,\ldots,c-1;\ g_1,\ldots,g_n = 0,1,\ldots,d-1\}$. S_{k+1} is obtained from $S_{k+1}{}^*$ by eliminating certain elements. Let $(\overline{s},\overline{t}) = (s_1,\ldots,s_n,t_1,\ldots,t_n)$ denote an element of $S_{k+1}{}^*$. $(\overline{s},\overline{t})$ is eliminated if $\sum_{i=1}^{n} a_ip^{s_i}q^{t_i} \not\equiv b \pmod r$ for some element r in P_{k+1}

which is not a power of p or q. Also, $(\overline{s},\overline{t})$ is eliminated if there is an element r in P_{k+1} such that r is a power of p or q and such that $\sum_{i\varepsilon T} a_i p^{s_i} q^{t_i} \not\equiv b \pmod{r}$ for each subset T of 1,2,...,n . Finally, $(\overline{s},\overline{t})$ is eliminated if the "other conditions" show that (*) has no solution $(\overline{x},\overline{y})$ with $\overline{x} \equiv \overline{s} \pmod{u_{k+1}}$ and $\overline{y} \equiv \overline{t} \pmod{v_{k+1}}$. (Also, symmetry assumptions may be employed to decrease the size of S_{k+1}).

The method by which S_0, S_1,..., S_ℓ are constructed assures the following: if $(\overline{x},\overline{y})$ is any solution of (*) and if $1 \leq k \leq \ell$, then S_k contains an element $(\overline{s},\overline{t})$ for which $\overline{x} \equiv \overline{s} \pmod{u_k}$ and $\overline{y} \equiv \overline{t} \pmod{v_k}$. In particular, S_ℓ contains an element $(\overline{s},\overline{t})$ such that $\overline{x} \equiv \overline{s} \pmod{\nu_p(S)}$ and $\overline{y} \equiv \overline{t} \pmod{\mu_p(S)}$.

Clearly $S \subseteq S_\ell$. In general, of course, the inclusion will be proper. However, if $S = S_\ell$, then the hypotheses of the Theorem are satisfied, and consequently S is the complete set of solutions of (*). The key question is: can we produce enough primes or prime powers in the sets P_1, P_2,...,P_ℓ so as to eliminate from the sets S_1, S_2,..., S_ℓ most (and eventually all) elements that do not correspond to actual solutions. No assertion is made that this can always be done, even in principle. However, the method works in many cases. The author has used it to solve all the degree equations for {2,3,5} and {2,3,7} groups in which some Sylow subgroup is cyclic and, as a consequence, to obtain a complete classification of simple groups of this type. In fact, if the "other conditions" of (*) are appropriate congruences, it seems practical (using a computer) to solve (*) whenever the number of terms does not exceed six or seven. In all cases investigated to date, the number of solutions has turned out to be finite. This raises the interesting theoretical question of whether there are general conditions under which (*) always has a finite number of solutions.

As an example, Table 2 gives the sequences $\{u_i\}$, $\{v_i\}$, and $\{P_i\}$ used to solve the equation (2). The congruences for $2^{x_i}3^{y_i}$ modulo 5 were used to eliminate elements from S_1*.

TABLE 1

SOLUTIONS OF EQUATION (2)

1) $1 + 1 + 1 + 1 - 4 = 0$

2) $1 + 1 + 1 - 9 + 6 = 0$

3) $1 + 1 - 4 + 6 - 4 = 0$

4) $1 + 1 + 16 + 6 - 24 = 0$

5) $1 + 1 + 16 - 54 + 36 = 0$

6) $1 + 1 + 16 - 9 - 9 = 0$

7) $1 - 9 - 4 + 6 + 6 = 0$

8) $1 - 9 - 4 + 16 - 4 = 0$

9) $1 - 9 - 4 + 36 - 24 = 0$

10) $1 - 9 + 16 + 16 - 24 = 0$

11) $1 - 9 - 64 - 9 + 81 = 0$

12) $1 - 9 - 64 + 36 + 36 = 0$

13) $1 - 9 - 64 + 96 - 24 = 0$

14) $1 - 9 - 64 - 144 + 216 = 0$

15) $1 + 81 - 4 - 54 - 24 = 0$

16) $1 + 81 - 64 + 6 - 24 = 0$

17) $1 + 81 - 64 - 54 + 36 = 0$

18) $1 - 729 - 64 + 576 + 216 = 0$

19) $1 - 729 + 256 + 256 + 216 = 0$

20) $1 - 729 - 1024 + 1536 + 216 = 0$

21) $1 - 729 - 16384 + 17496 - 384 = 0$

TABLE 2

i	u_i	v_i	P_i
0	1	1	
1	2^2	2^2	2^4,3
2	$2^2 3$	$2^2 3$	7, 13, 3^2
3	$2^2 3^2$	$2^2 3$	73
4	$2^2 3^2$	$2^2 3^2$	19, 37
5	$2^3 3^2$	$2^4 3^2$	17, 2^6
6	$2^4 3^2$	$2^4 3^2$	97, 577
7	$2^4 3^2$	$2^4 3^3$	109, 433
8	$2^4 3^4$	$2^4 3^3$	3^5, 2^6
9	$2^4 3^4$	$2^4 3^4$	163, 1297, 3889
10	$2^4 3^5$	$2^4 3^5$	487
11	$2^4 3^5$	$2^4 3^6$	1459, 2917
12	$2^4 3^7$	$2^4 3^6$	17497, 2^6, 3^8

University of Illinois, Chicago Circle

ON FINITE LINEAR GROUPS OF DEGREE LESS THAN $(q - 1)/2$

Henry S. Leonard, Jr.

Let G be a finite group, and let P be a Sylow subgroup of G with normalizer N and centralizer C. Then the following can be proved.

THEOREM. *Suppose the centralizer of every non-identity element of P is C, and assume N/C is abelian. If G has a faithful complex representation of degree less than $(|P| - 1)/2$ then P is normal in G.*

This is a variation of [2], theorem 1 and is a generalization of a theorem of Brauer [1], theorem 3, which provides the above result when P has prime order.

The proof is based on a variation of [2], theorem 2, and on [2], theorem 3, and provides a new proof of Brauer's theorem except that it is convenient to apply his theorem when $|P| = 7$ or 13. Generally the methods used are an extension of those employed in [2]. The details will appear elsewhere.

REFERENCES

[1] R. Brauer, On groups whose order contains a prime number to the first power, II, Amer. J. Math. 64 (1942), 421-440.

[2] H. S. Leonard, Jr., Finite linear groups having an abelian Sylow subgroup, J. Algebra 20 (1972), 57-69.

Northern Illinois University

A SETTING FOR THE LEECH LATTICE

John McKay

Let

$$B_d = (k+1)^{-1/2} \left[\begin{array}{c|c} (k+1)I_n & H_n - I_n \\ \hline O_n & I_n \end{array} \right]$$

where $d = 2n = 8k$ and H_n (=H) is a skew Radamard matrix of order n, i.e., $H = -I+S$ is a matrix with entries $\pm$ 1 such that $H^tH = nI$ and $S = -S^t$.

Over $\mathbb{Z}$ the columns of B_d generate a unimodular lattice Λ_d which is even for odd values of k and has minimum 4 for such $k \neq 1$. It follows that Λ_8 is the lattice associated with the Weyl group of type E_8 and that Λ_{24} is the Leech lattice.

Details will appear elsewhere.

REFERENCES

[1] E. F. Assmus and H. F. Mattson, Perfect codes and the Mathieu groups, Arch. Math. 17 (1966), 121-135.

[2] J. H. Conway. A characterization of Leech's lattice, Invent. Math. 7 (1969), 137-142.

[3] W. M. Kantor, Automorphism groups of Hadamard matrices, J. Comb. Theory 6 (1969), 279-281.

[4] J-P. Serre, Cours D'arithmetique. Presses Universitaires

de France (1970).

School of Computer Science
McGill University

THE NORMAL STRUCTURE OF THE ONE-POINT STABILIZER OF A DOUBLY-TRANSITIVE GROUP

Michael O'Nan

We use the following notation. G is a doubly-transitive group on a set X and G_x is the stabilizer of $x \in X$. If N^x is a normal subgroup of G_x, we define N^y so that $fN^y f^{-1} = N^{f(y)}$. Thus, N^y is the unique conjugate of N^x which lies in G_y.

Most of the theorems we state began as an attempt to extend the following theorem.

THEOREM A. *If G is a doubly-transitive permutation group on X and A^x is a normal subgroup of G_x such that A^x is abelian and not semi-regular on X-x, then* $PSL(n,q) \leq G \leq P\Gamma L(n,q)$, $n \geq 3$.

In this case, it turns out that A^x is the group of transvections with center x.

Early in the proof of this theorem we establish the condition $A^x \cap A^y = 1$ if $x \neq y$, or in other words, that A^x is a T.I. set. Moreover, it becomes apparent that this condition plays a central role in many of the arguments.

Accordingly, using similar methods, but a different point of view, we are led to Theorem B.

THEOREM B. *If G is doubly-transitive on X and* $N^x \triangleleft G_x$, *with*

(i) $N^x \cap N^y = 1$, if $x \neq y$ (i.e., N^x *is a* T.I. *set*)

(ii) N^x *is not semi-regular on* X-x

then $PSL(n,q) \leq G \leq P\Gamma L(n,q)$.

This theorem has several consequences, the first of which is of a geometrical nature.

COROLLARY 1. *Let* B *be a block design on* X (*with* $\lambda = 1$). *Suppose that* Aut(B) *is doubly-transitive on* X. *Let* N^x *be the subgroup of* Aut(B) *which fixes all blocks containing* x. *Assume* $N^x \neq 1$ *and* N^x *is not semi-regular on* X-x.

Then, B *is a Desarguesian projective geometry.*

Another consequence yields information about the normal structure of the one-point stabilizer of a doubly-transitive group.

COROLLARY 2. *Let* G *be a doubly-transitive group on* X. *Then either*

(1) G_x *is a local group, or*

(2) *The minimal normal subgroup of* G_x *is semi-simple, self-centralizing and restricts faithfully to each of its orbits.*

The third consequence of this theorem is of an arithmetic nature. Essentially, it bounds the degree of a doubly-transitive group in terms of the length of the orbits of a normal subgroup (when such a bound is possible).

COROLLARY 3. *Let* G *be a doubly-transitive group on a set* X, *with* $|X| = n$. *Let* N^x *be a normal subgroup of* G_x *and suppose that the orbits of* N^x *are of length* ℓ *on* X-x. *Then, either*

(i) N^x *is semi-regular on* X-x, *or*

(ii) $PSL(n,q) \leq G \leq P\Gamma L(n,q)$, *or*

(iii) $n \leq 2\ell^2$.

In the first two cases there is no hope of bounding the degree in terms of ℓ. For example, if p is an odd prime, E an elementary abelian p-group, and G the semi-direct product of E and its full group of automorphisms, then G is a 2-transitive group of degree $|E|$ and G_x has a normal subgroup having orbits of length 2. In this case N^x is semi-regular. Likewise, in PSL(n,q), no such bound is possible.

This does not appear to be the best possible bound. It seems likely that the bound $n \leq (\ell - 1)^2$ holds.

While the corollaries cannot be obtained immediately from Theorem B, the bulk of the work likes in the proof of Theorem B.

Central to this proof is the study of the following objects. By an (H,K,L) configuration we mean that H,K,L are groups, $H \leq Aut(K)$ and $L \leq K$, and for all $x \in H-1$, $C_K(x) = L$.

It is not difficult to show that if $N^x \triangleleft G_x$ and $N^x \cap N^y = 1$, we obtain an (N_x^y, N^x, N_y^x) configuration. Moreover, in this case, $N_x^y \cong N_y^x$.

Accordingly, in proving theorem B we study (H,K,L) configurations in which $L \cong H$. We prove that under these circumstances either

(1) H is abelian or

(2) H is a Frobenius complement, or

(3) Both H and K are Frobenius groups.

The methods in proving this are more or less standard local group theoretic arguments.

Once we have obtained precise enough information concerning the structure of N^x, geometrical arguments are used to prove that the group is PSL(n,q), or that a given configuration does not occur.

Rutgers University

FLAG-TRANSITIVE SUBGROUPS OF CHEVALLEY GROUPS

Gary M. Seitz

Let G be a finite group of Chevalley type and B the Borel subgroup of G. If $L \leq G$ and $G = BL$, then L is said to be flag-transitive. In [3] D. Higman determined all flag-transitive subgroups of G in case G is of type A_n. The main result here is an extension of Higman's theorem to the case of a general group of Chevalley type. The result is then used to show the nonexistence of 2-transitive permutation representations of certain groups of Chevalley type.

Let G_0 be a Chevalley group of normal or twisted type such that $Z(G_0) = 1$. Let $G_0 \leq G_1 \leq Aut(G_0)$ where G_1 is generated by G_0 and the diagonal and field automorphisms of G_0. We say that G is of Chevalley type provided $G_0 \leq G \leq G_1$. If G is of Chevalley type and B the Borel subgroup of G, let $U = Fit(B)$.

THEOREM 1. *Let G be a group of Chevalley type of rank $n \geq 2$ and suppose $L \leq G$ is flag-transitive. Then either $U^G \leq L$ or one of the following holds:*

i) $G \cong PSL(3,2)$ *and* $|L| = 3 \cdot 7$.

ii) $G \cong P\Gamma L(3,8)$ and $|L| = 3^2 \cdot 73$.

iii) $G \cong PSL(4,2) \cong A_8$ *and* $L \cong A_7$.

iv) $G \cong Sp(4,2) \cong S_6$ *and* $L \cong A_6$.

v) $G \cong G_2(2)$ *and* $L = G'$.

vi) $G \cong {}^2F_4(2)$ *and* $L = G'$.

vii) $G \cong PSp(4,3) \cong PSU(4,2)$ *and L is a maximal parabolic subgroup of* $PSU(4,2)$ *of order* $2^6 \cdot 3 \cdot 5$.

If G is a group of Chevalley type, we write G = G(q) to indicate that G is defined over the field $\mathbb{F}_q$. As application of Theorem 1 we have the following results.

THEOREM 2. _Let_ W = $\langle s_1,\ldots,s_n\rangle$, $n \geq 2$ _be an indecomposable coxeter group. There is an integer_ N, _depending on_ W, _so that if_ G = G(q) _is a group of Chevalley type with_ $q \geq N$ _and having Weyl group of type_ W, _then_ G _has no faithful_ 2-_transitive permutation representation unless_ W _is of type_ A_n _and_ G _is represented on the cosets of a maximal parabolic subgroup of_ G.

THEOREM 3. _Let_ G _be a group of Chevalley type with_ $U^G \cong PSp(4,q)$, $q > 2$, $PSU(4,q)$, $PSU(5,q)$, $G_2(q)$, $q > 2$, _or_ ${}^3D_4(q)$. _Then_ G _has no non-trivial_ 2-_transitive permutation representation_.

The connection between Theorems 2 and 3 and Theorem 1 can easily be seen as follows. Let G = G(q) be a group of Chevalley type with $q = p^a$ and p prime. Suppose that G is 2-transitive on the cosets of $L \leq G$. Then $1_L^G = 1 + \theta$ for θ an irreducible character of G. If $p|\theta(1)$ then $|G:L|$ is prime to p and L contains a Sylow p-subgroup of G. In this case it easily follows that L is a maximal parabolic subgroup and consequently G is of type A_n. Suppose $p \nmid \theta(1)$. Write $1_B^G = 1 + \chi_1 + \ldots + \chi_r$ when $\chi_1,\ldots,\chi_r$ are irreducible characters of G. If we know that $p \mid \chi_i(1)$ for each $i = 1,\ldots,r$ then $1 = (1_L^G, 1_B^G)$ = the number of L,B-double cosets of G. Thus G = BL and L is flag-transitive. Thus the only remaining question is whether or not $p \mid \chi_i(1)$ for each $i = 1,\ldots,r$. By applying a result of Green [2] we obtain Theorem 2. Theorem 3 follows from work of Curtis, Iwahori, and Kilmoyer [1].

From the above it is clear that Theorem 1 can be used to determine all 2-transitive permutation representations of all the groups of Chevalley type once it is known that with only a few exceptions p divides the degree of all non-principal irreducible constituents of $1_B{}^G$ for G a group of Chevalley type and of characteristic p.

The proof of Theorem 1 proceeds as follows. Let p be the characteristic of G, so that U is a p-group. The basic aim is to show that a large portion of U is contained in L. In particular we show that with few exceptions $L \cap P$ is 2-transitive on the cosets of B in P, where P is a parabolic subgroup of G generated by B and one of the fundamental reflections generating the Weyl group of G. Once this is done for each such P, the proof is completed rather easily.

It is easy to see that we need only consider the case where $|G:U^G|$ is prime to p. The first troublesome point is the possibility that $L \cap U = 1$ and L is a p'-group. Here the idea is to find a low dimensional (projective) representation of U^G and then lift this to a complex representation of L. Then finding large prime divisors of $|L|$ we are able to apply results of Feit to obtain information about the structure of L, and this leads to a contradiction. Next we separately handle the case $G = G(q)$ and $q = 2$. Using an inductive lemma regarding the rank of G we can reduce to case of a group of rank at most 4, with the exception of Sp(2n,2). After giving a separate argument to eliminate Sp(2n,2) with $n \geq 3$ we use order considerations for the remaining cases. throughout this section of the paper and later ones detailed information concerning the structure of U is used when G is of rank 2.

Next we assume $q > 2$, $L \cap U \neq 1$, and show that there is at least one 2-transitive section as described above. Once this is accomplished we use the fact that the Dynkin diagram is connected, reduce to the case of a rank 2 group, and show all the appropriate sections are 2-transitive.

REFERENCES

[1] C. W. Curtis, N. Iwahori, and R. Kilmoyer, Hecke algebras and characters of parabolic type of finite groups with (B,N)-pairs, Math. Pub. I.H.E.S., (to appear).

[2] J. A. Green, On the Steinberg character of finite Chevalley groups, Math. Zeit. 117 (1970), 272-288.

[3] D. G. Higman, Flag-transitive collineation groups of finite projective spaces, Ill. J. Math. 6 (1962), 434-446.

University of Oregon

COROLLARIES OF STRONGLY EMBEDDED TYPE FROM A THEOREM OF ASCHBACHER

Raymond Shepherd and Ernest Shult

Recently Aschbacher has classified groups containing a proper 2-regular core [1]. In the course of proving this important result, he also proves the following fundamental theorem:

THEOREM. *Let* z *be an involution lying in a proper subgroup* H *of a finite group* G. *Suppose* $|z^G|$ *is odd and*

(i) $z \in g^{-1}Hg$ *for* $g \in G$, *implies* $g \in H$

(ii) *If* u *is an involution such that*

$$z \in C_G(u) \not\leq H$$

then $H \cap \langle C_G(u) \cap z^G\rangle$ *is strongly embedded in* $\langle C_G(u) \cap z^G\rangle$.

Under these conditions $\langle H \cap z^G\rangle$ *is strongly embedded in* $\langle z^G\rangle$.

The first observation is

COROLLARY 1. *The Aschbacher Theorem remains valid if conditon* (ii) *is replaced by*

(ii)' *If* u *is an involution such that* $z \in C_G(u) \not\leq H$, *then* $u \notin \langle C_G(u) \cap z^G\rangle$.

A useful feature of this corollary is that the isomorphism type of the weak closure of z in $C_G(u)$ is not specified and hence can be applied to a variety of theorems whose conclusion is the assertion that $\langle z^G\rangle$ belongs to a class of groups closed under normal products.

The proof of the corollary is a straight-forward induction. One need only show that if the involution u appearing in (ii)' is central in G, then hypothesis (i) and (ii)' hold for $\overline{z} = z\langle u\rangle$ and $\overline{H} = H/\langle u\rangle$ in the roles of u and z. (The new involutions $\overline{u}$ entering the statement of (ii)' for $\overline{G} = G/\langle u\rangle$ may be images of elements of order 4 in G.)

The second corollary of Aschbacher's Theorem represents a far-reaching generalization of Bender's Theorem [2] on strongly embedded subgroups.

COROLLARY 2. *Let J denote the set of involutions in a subgroup H of G. Suppose no involution in J commutes with an involution in $J^G - J$. Then either $J = J^G$ or $N(J) \cap \langle J^G\rangle$ is a strongly embedded subgroup of $\langle J^G\rangle$*

PROOF. We may assume $J^G \neq J$; in particular J is not empty.

Let C be the graph whose vertices are the elements of J^G and whose arcs are commuting pairs of involutions in J^G. Then J is a union of connected components $J_1, J_2, \ldots, J_k$ of C. First we show J may be assumed to be connected.

Suppose J were not connected. Then J clearly consists of one H-class of involutions as k is odd and $|J_1| = |J_2| = \ldots = |J_k|$ is odd. Suppose $x \in J_r$. If x stabilized J_s, $s \neq r$, under conjugation then x would centralize an involution in J_s since $|J_s|$ is odd, contrary to the definition of the J_s's. Thus x stabilizes only J_r. Now set $H_1 = \langle J_1\rangle$. Then $H_1 \leq \langle J\rangle \leq H$ and any involution t in H_1 lies in J and hence in some J_r. But since all involutions in J_1 normalize J_1, we have $H_1 \leq N(J_1)$ and so $t \in J_r \cap N(J_1)$. But we have just observed that this means $t \in J_1$. Thus J_1 is also the set of *all* involutinns in $H_1 = \langle J_1\rangle$. Now the hypotheses of the theorem hold with H_1 replacing H. Induction on $|J|$ and the fact that $J_1^G = J^G$ yields the result.

Thus we may assume that J is itself a connected component of $\mathcal{C}$. In particular J is a trivial intersection set in G. Since any element of $J^G \cap N(J)$ must centralize an element of J, $J^G \cap N(J) \leq J$. Let S_o be a 2-Sylow subgroup of K; let S_1 be a 2-Sylow subgroup G containing S_o, and set $S_2 = N_{S_1}(S_o)$. Then, since J is non-empty, also $J \cap S_o$ is non-empty. If $y \in S_2 - S_1$, $(J \cap S_o)^y \subseteq J^G \cap S_0 \subseteq J^G \cap N(J) \subseteq J$. Since J is a trivial intersection set, this implies $y \in N(J) \cap S_2 = S_0$, contrary to our choice of y. Thus $S_0 = S_1 = S_2$ is a 2-Sylow subgroup of G.

Set $K = N_G(J)$ and let $z \in J$ belong to the center of S_0. If $z \in K^g$ for some g in G, z normalizes J^g and so commutes with an element in it, since $|J^g|$ is odd. Thus $J = J^g$, being a connected component of $\mathcal{C}$, and so $g \in K$. Thus we have verified that condition (i) of Aschbacher's Theorem holds with our K in the role of his H and z an involution in J, chosen to lie in a 2-Sylow center of G. (Note that $J^G \neq J$ makes K a proper subgroup of G.)

Suppose u is an involution in K and $C_G(u) = G$. Then as J^G is not connected, $u \notin J$, and so $u \notin H$. Thus every involution in $\overline{H} = (H \times \langle u\rangle)/\langle u\rangle$ is an image of an element in J. Moreover, $Ju \cap J^G$ is empty. It follows that for $x \in J$, $N_G(\langle x,u\rangle) \leq C_G(x) \leq K$. Thus $\overline{J} = J\langle u\rangle$, $\overline{H} = H\langle u\rangle/\langle u\rangle$ and $\overline{G} = G/\langle u\rangle^G$ satisfy the hypotheses of Corollary 2. Induction and $\overline{J} \neq \overline{J}^G$ yield $(\langle J\rangle \times \langle u\rangle)/\langle u\rangle$ is strongly embedded in $\langle J^G\rangle\langle u\rangle/\langle u\rangle$, and the conclusion of the theorem follows easily from this. Thus we may assume for each involution u in K, that $C_G(u)$ is a proper subgroup of G.

If L is a proper subgroup of G which meets J non-trivially, form $H_L = \langle J \cap L\rangle$. Then $J \cap L$ consists of all involutions in H_L, and no involution in $J \cap L$ commutes with an involution in $(J \cap L)^L - (J \cap L)$. Thus induction on $|L|$ yields

$(J \cap L)^L = J \cap L$ or $\langle J \cap L\rangle$ strongly embedded in $\langle (J \cap L)^L\rangle$. In the former case, since $J \cap L$ is non-empty and J is a T.I. set, $L \leq K$.

Fix z in J so that z lies in a 2-Sylow subgroup of G. Let u be an involution in K such that $z \in C(u) \not\leq K$. From the above, $C(u) < G$, and so $\langle J \cap C(u)\rangle^{C(u)}$ is strongly embedded in $\langle (J \cap C(u))^{C(u)}\rangle$. If $z^x \in C(u)$, then z^x normalizes $\langle (J \cap C(u))^{C(u)}\rangle$ and (since the latter has one class of involutions) one can find $c \in C(u)$ such that z^{xc} commutes with an involution in $J \cap C(u)$. Since J is a connected component of $\mathcal{C}$, this places $z^{xc} \in J$. Thus

$$\langle x^G \cap C(u)\rangle \leq \langle (J \cap C(u))^{C(u)}\rangle$$

and since the right hand side contains one class of involutions, equality holds.

Now all of the hypotheses of Aschbacher's theorem are operating and so $\langle z^G\rangle$ contains $K \cap \langle z^G\rangle$ as a strongly embedded subgroup.

Thus if $W = \langle z^G\rangle$, $W/O(W) \cong Z_2$ or is a simple Bender group.

It remains only to show that $J^G = J$ or else $\langle z^G\rangle = \langle J^G\rangle$. Deny the latter and suppose $t \in J - W$. If $W/O(W) = \overline{W}$ is a simple Bender group, t either induces a field automorphism or an inner automorphism on $\overline{W}$. In either case there exists an element $w \in z^G \cap C(w) \subseteq J$ such that $t' = t$ or tw centralizes involutions belonging to at least 2 distinct 2-Sylow centers, $\overline{T}_1$ and $\overline{T}_2$ of $\overline{W}$. Since O(W) has odd order, we may assume $\overline{T}_i = T_iO(W)/O(W)$ where T_i is elementary and meets $C(t')$ non-trivially. It follows that $T_i^\# \subseteq J$, and since T_i contains a 4-group, $(T_i^\#)^{O(W)} \subseteq J$, $i = 1,2$. But the preimage in W of the normalizer of $\overline{T}_1$ then lies in K, and transitively permutes the sets $(T_2^\#)^{O(W)x}$, $x \in W$, distinct from

$(T_1^{\#})^{O(W)}$. It follows that every involution in W belongs to J. Since every involution in J^G commutes with at least one involution in W, and since J is a connected component of $\mathcal{C}$, this yields $J^G = J$. Thus one is left with the case $W/O(W) \cong Z_2$ (so $z \in \mathbb{Z}^*(G)$). Then find $z_1 \in W \cap C(t)$, so $\langle t, z_1 \rangle$ is a fours-subgroup of H. Then $(\langle t, z_1 \rangle^{\#})^{O(W)} \subseteq J$ so (since J is a trivial intersection set) $O(W) \leq N(J)$. But $C(z_1) \leq N(J)$ and so $G = O(W)C_G(z_1) \leq N(J)$. This means $J^G = J$ and completes the proof.

An application of Corollary 2 appears in [3] of these Proceedings.

REFERENCES

[1] M. Aschbacher, Finite groups with a proper 2-generated core. To appear.

[2] H. Bender, Transitive Gruppen gerader Ordnung in denen jede Involution genau einen Punkt festält, J. Alg. 17 (1971), 527-554.

[3] E. Shult, On fusion in 2-Sylow intersections. These proceedings, page 131.

Elmhurst College and University of Florida

University of Florida

ON FUSION IN 2-SYLOW INTERSECTIONS

Ernest Shult

The very least that one might say about fusion in proper 2-Sylow intersections should entail a description of what occurs when, in some respect, there isn't any fusion. Accordingly, the theorem described below concerns the following hypothesis:

(A) *Let z be an involution which lies in the center of some 2-Sylow subgroup of a finite group G. Suppose z is isolated in every proper 2-Sylow intersection which contains it; that is, if S and S^g are two distinct 2-Sylow subgroups of G, then* $|z^G \cap S \cap S^g| \leq 1$.

This hypothesis occurs in various places. For example, M. Herzog [7] considered groups G containing an involution lying in the center of a 2-Sylow subgroup, but which lies in no proper 2-Sylow intersections of G. Groups in which all proper 2-Sylow intersections are rank 1 have been studied by Mazurow [14] (for cyclic intersections) and have been classified completely by M. Aschbacher [1] and later independently by P. Landrock [1]]. Even this situation can be generalized in one direction. Let an involution z be called a *central involution* if it lies in the center of some 2-Sylow subgroup of a group G. A Sylow intersection is called *central* if it is a proper Sylow intersection which is either order one or contains a central involution. In a series of short papers by Herzog [8,9] and Herzog and Shult [10], simple groups whose central 2-Sylow intersections have 2-rank one are classified.

Under any of the assumptions considered in the previous paragraph, any central involution z in G satisfies the hypothesis on isolation (A). The theorem presented below was an attempt to generalize the results quoted above.

THEOREM. Let the involution z and the group G satisfy Hypothesis (A). Let S denote any 2-Sylow-subgroups of G containing z and let $N = \langle z^G \rangle$, the smallest normal subgroup of G containing z. Then one of the following holds:

(a) z lies in no proper 2-Sylow intersection and N is a central product of perfect odd central extensions of Bender groups and a 2-nilpotent group with elementary 2-Sylow subgroup E such that $[E: E \cap O_2(N)] \leq 2$.

(b) $N \simeq J$, Janko's simple group of order 175,560.

(c) $O_{2'}(N) \neq \langle 1 \rangle$, $N/O_{2'}(N) \simeq SL(2,4)$ or $U(3,4)$.

(d) $N/O_{2'}(N) \simeq Z_2$, i.e. $z \in \mathbb{Z}^*(G)$.

(d') $N/O_{2'}(N)$ is an elementary 2-group of order ≥ 4, and z lies in a proper 2-Sylow intersection.

(e) $N/O_{2'}(N) \simeq PSL(2,q)$, $q \equiv 3$ or $5 \bmod 8$.

(f) The weak closure $T = \langle z^G \cap S \rangle$ of z in S is a strongly closed fours-subgroup of the center of S. Involutions in $S - T$ exist and for any such involution, x, some involution $t \in xT$ has the property that $\langle z^G \cap C(t) \rangle$ is a semidirect product of $PSL(2,q)$ ($q \equiv 3$ or $5 \bmod 8$) and a normal subgroup of odd order.

I call this an "attempt" at a generalization since (f) represents a shortcoming of the theorem. Ideally, case (f) should be reducible to cases (d') or (e), so that (a) - (e) represents the entire conclusion of the theorem. Nonetheless case (f) involves a very small weak closure, so it is hoped that the theorem may still prove useful in determining the possible patterns of fusion in 2-Sylow subgroups of 2-simple groups.*

We conclude with a few remarks concerning the proof of the theorem. By Glauberman's Z*-theorem [3], we may assume z is not isolated in $C_G(z)$ since otherwise case (d) holds. Hypothesis (A) implies that z lies in the center of each 2-Sylow subgroup containing it. Then for each conjugate z^g in $C_G(z) - \langle z\rangle$, $z^g z = \overline{z}^g$ is a central involution of $\overline{C} = C_G(z)/\langle z\rangle$ lying in no proper 2-Sylow intersection of $\overline{C}$. Then the normal closure $\Theta(z^g)$ of $\overline{z}^g$ in $\overline{C}$ has the form given for N in case (a) by [7]. As z^g ranges over $z^G \cap C(z) - \{z\}$, one obtains a normal product $\overline{W}$ of various $\Theta(z^g)$'s, which also has the form described for N in case (a). (This is Theorem 1.9 in [16]. Then $\overline{W} = W/\langle z\rangle$ where $W = \langle z^G \cap C(z)\rangle$. Setting $T = \langle z^G \cap S\rangle$, we see that $\overline{T} = T/\langle z\rangle$ is a 2-Sylow center of $\overline{W}$ and T is an elementary subgroup of the center of S. Each $\overline{z}^g$ is an involution of $\overline{W}$ which projects non-trivially on each of the simple or 2-nilpotent components of $\overline{W}/Z(\overline{W})$. This means that on the set $z^G \cap T$, all fusion is effected by $B = N(T)/C(T)$ having odd order and for each quasisimple component $\overline{X}$ of $\overline{W}$, there exists a cyclic subgroup $\overline{Q}_X$ in $\overline{X} \cap (N(T) \cap W)/\langle z\rangle$ fixing z but acting as a <u>semiregular group of permutations</u> on $z^G \cap T - \{z\}$ (the elements project non-trivially on $\overline{X}$). Since B is a transitive solvable group of permutations on $z^G \cap T$, $\overline{Q}_X$ must give Frobenius action on some non-trivial subgroup of F(B), the Fitting subgroup of B. On the

---

*It appears that case (f) is totally eliminated by the imminent theorem of Goldschmidt on strongly closed abelian subgroups of a 2-Sylow subgroup.[5] Indeed, almost all of the arguments of this paper will be subhumed at once under Goldschmidt's important result.

$\overline{Q}_X$ acts as a linear group on T, regularly permuting the elements of $[T,\overline{Q}_X]^{\#}$. This contrast of viewpoints forces $\overline{Q}_X \simeq Z_3$. If two quasisimple simple components X and Y are involved in W, then $\overline{Q}_X \times \overline{Q}_Y \simeq Z_3 \times Z_3$ must act semiregularly on $z^G \cap T - \{z\}$, and so gives Frobenius action on a suitable section of F(B), namely $N/(N \cap \ker Z_1)$ where N is a minimal normal subgroup of B, Z_1 is a N-orbit containing z and $\ker Z_1 \cap N$ is the kernel of the action of N on Z_1. Since Frobenius action by $Z_3 \times Z_3$ is clearly impossible, at most one quasisimple component is involved in $\overline{W}$. Two cases arise: $\langle z^G \cap T\rangle = T = [T,N]$ and $z^G \cap T \subseteq T - (C_T(N) \cup [T,N])$. In the latter case $|T| = 2^4$, $\overline{Q}_X N$ is metacyclic of order 21 and yields a fusion pattern 1-7-7. This is also the B-fusion pattern since the group of order 21 is a maximal odd-order subgroup of L(4,2). Then T contains an involution t isolated in S (since N(T) controls fusion for this). This involution t is not conjugate to z, and lies in Z*(G). Induction to G/Z*(G) easily gives one of the results (a) - (f) for G. In the case $T = [N,T]$, $N \simeq Z_7$, and $W \simeq Z_2 \times U(3,4)$ (easily disposed of) or $W \simeq Z_2 \times SL(2,4)$. In the latter case, (b) holds by a result of Shult [17].

Thus for each $z \in z^G$, $\overline{W}_z = W_z/\langle z\rangle$ has no quasisimple components where $W_z = \langle z^G \cap C(z)\rangle$. Indeed, C(z) is solvable, with 2-length one.

Suppose u is an involution in $S - z^G$, and $|T| > 4$. An easy induction yields $O_{2'}(G) = 1$ and $u \notin Z(G)$. Then T is in a 2-Sylow center of $W_u = \langle z^G \cap C(u)\rangle$, a normal product of groups $N_{u,i} = \langle z_i^{C(u)}\rangle$ of the shape (a) - (f). Suppose W_u possesses a non-solvable composition factor not isomorphic to J or a Bender group. Then one of the $N_{u,i}$ contains such a factor, say $N_{u,1}$. Then $N_{u,1}$ is type (e) and is a perfect group. Since $|T| > 4$, and $N_{u,1}$ has a

2-Sylow subgroup of order 4, there exists an element $t \in z^G \cap C(u) - N_{u,1}$. Say $\langle t^{C(u)} \rangle = N_{u,2}$. Then $[t, N_{u,1}] \leq N_{u,1} \cap N_{u,2}$. If $N_{u,1} \cap N_{u,2} < N_{u,2}$, then $N_{u,1} \cap N_{u,2} \leq O(N_{u,1})$ since $N_{u,1}/O(N_{u,1}) \simeq PSL(2,q)$, q odd > 3, is simple. Thus C(t) covers this section, and this contradicts the fact which we learned from the previous paragraph, that C(t) is solvable. Thus $N_{u,1} \leq N_{u,2}$ and so $N_{u,2}$ has PSL(2,q) as a composition factor. Symmetry now yields $N_{u,1} = N_{u,2}$ against our choice of t. Thus if $|T| \geq 8$, W involves only J or Bender groups as composition factors, and since $T \leq W_u$ implies $C(u)/W_u$ is 2-closed, we have that C(u) is 2-constrained. Since this holds for all $u \in T$, $O_{2'}$ is a T-signalizer functor. Then, as T has rank ≥ 3, by the theorem of Goldschmidt [4],

$$M = \langle O_{2'}(C(t)) \mid t \in T^{\#} \rangle$$

has odd order and is normalized by N(T) = N(S). Since $O_{2'}(G) = 1$, L = N(T)M is a proper subgroup of G and $\langle z^G \cap L \rangle = TM = H$. Every involution u in T has W_u solvable so $C(u) = (C(u) = (C(u) \cap N(T))O(C(u)) \leq L$. Since N(T) controls all 2-fusion in itself,

$$u^G \cap C(u) \leq H = TM \text{ for every involution in } H.$$

By corollary 2 of [15], (a corollary of Theorem 2 of Aschbacher [2]) $\langle x^G \rangle \leq H$ or else $H \cap \langle z^G \rangle$ is strongly embedded in $\langle z^G \rangle$. The first case yields (d') and the second yields (a) for G.

Thus we may assume $|T| = 4$. Then as $|z^G \cap T|$ is odd, $T^{\#}$ is fused, and lies in the center of S. If S - T contains no involutions, S contains just three involutions and cases (c). (d), (d'), or (e) hold for G by the theorems of Gorenstein and Walter [6], Higman [11], and Lyons [13]. Let s be an involution in S - T. Then $A = \langle s, T \rangle$ is elementary of order 8. Suppose for each involution

$a \in A^{\#}$, that $C(a)$ was 2-constrained. Then the signalizer functor theory applies and so $M = \langle O_{2'}(C(t)) | t \in T^{\#} \rangle$ has odd order (being a subgroup of $\langle O_{2'}(C(a)) | a \in A \rangle$). Then $L = N(T)M$ leads to a strongly embedded subgroup or case (d') by means of Corollary 2 of [15] in these proceedings. Thus $C(a) = (C(a) \cap N(T))W_a$ and hence $W_a = \langle z^G \cap C(a) \rangle$ is not 2-constrained for some $a \in A - T$. Then induction applies to $C(a)$, and since W_a is not 2-constrained, case (e) or (f) holds for $C(a)$. In case (e), $W_a/O(W_a) \simeq PSL(2,q)$ and in case (f), $\langle z^{C(a)} \cap C(b) \rangle \simeq PSL(2,q)$ acting on a group of odd order, for some b in $aT = A - T$. Thus in either case (f) holds for G, completing the proof.

REFERENCES

[1] M. Aschbacher, A class of generalized TI groups. To appear in Ill. J. Math.

[2] __________, Finite groups with a proper 2-regular core.

[3] G. Glauberman, Central elements in core free groups, J. Alg. 4 (1966), 403-420.

[4] D. Goldschmidt, 2-Signalizer functors on finite groups, J. Alg. 20 (1972), 333 -

[5] __________, Strongly closed abelian 2-subgroups of finite groups. These Proceedings, p. 55.

[6] D. Gorenstein and J. Walter, The characterization of finite groups with dihedral Sylow 2-subgroups, I, II, III, J. Alg. 2 (1965), 85-151, 218-270, 334-393.

[7] M. Herzog, On 2-Sylow intersections, to appear in Math. Zeit.

[8] __________, Central 2-Sylow intersections, to appear.

[9] __________, Simple groups with cyclic central 2-Sylow

intersections, to appear.

[10] M. Herzog and E. Shult, Groups with central 2-Sylow intersections of rank at most one, to appear.

[11] G. Higman, Suzuki 2-groups, Illinois J. Math. 7 (1963), 79-96.

[12] P. Landrock, Finite groups with Sylow 2-intersections of rank $\leq$ 1. Aarhus University preprint series #23. To appear elsewhere.

[13] R. Lyons, A characterization of $U_3(4)$, to appear.

[14] V. D. Mazurow, Finite simple groups with cyclic intersections of Sylow 2-subgroups, Algebra i Logika 10 (1971), 188-198, in Russian.

[15] R. Shepherd and E. Shult, Corollaries of strongly embedded type from a theorem of Aschbacher. These proceedings, 126-130.

[16] E. Shult, On the fusion of an involution in its centralizer, to appear.

[17] __________, A note on Janko's simple group of order 175,560, to appear in Proc. Amer. Math. Soc.

University of Florida

THE EXISTENCE AND UNIQUENESS OF LYONS' GROUP

Charles C. Sims

In [1] Lyons considered a finite simple group G possessing an involution whose centralizer in G is isomorphic to $\tilde{A}_{11}$, the two-fold covering group of A_{11}. He was able to determine the character table and much of the structure of G. In particular, he showed that the order of G is

$$51765179004000000 = 2^8 \cdot 3^7 \cdot 5^6 \cdot 7 \cdot 11 \cdot 31 \cdot 37 \cdot 67$$

and that G contains a subgroup H isomorphic to $G_2(5)$. The index of H in G is 8835156.

The author has shown that a group G satisfying Lyons' hypothesis exists and is unique up to isomorphism. The complete proof of this result involves approximately 100 pages of text and 17 computer programs which total over 2000 lines of FORTRAN. The combined running time for these programs is about one hour on an IBM 360/67, roughly $400 worth of time at any major university computing center. All that will be given here is a presentation for G. This presentation appears somewhat formidable, but computer programs exist for solving the related word problem very efficiently. The details of the proofs of existence and uniqueness together with the computer programs and the data necessary to solve the word problem in G will be published in [2].

Consider the following 7 x 7 matrices over GF(5):

$$a = \begin{bmatrix} 4&0&0&0&0&0&0 \\ 0&0&0&0&3&0&0 \\ 0&0&0&2&0&0&0 \\ 0&0&1&0&0&0&0 \\ 0&1&0&0&0&0&0 \\ 0&0&0&0&0&0&2 \\ 0&0&0&0&0&3&0 \end{bmatrix}, \quad b = \begin{bmatrix} 4&0&0&0&0&0&4 \\ 0&2&0&0&2&0&0 \\ 0&0&0&3&0&0&0 \\ 0&0&3&2&0&0&0 \\ 0&2&0&0&0&0&0 \\ 0&0&0&0&0&0&1 \\ 3&0&0&0&0&1&4 \end{bmatrix},$$

$$c = \begin{bmatrix} 1&1&0&0&2&0&0 \\ 0&1&0&0&0&0&0 \\ 3&1&1&0&3&2&4 \\ 1&3&0&1&4&3&4 \\ 0&0&0&0&1&0&0 \\ 0&1&0&0&1&1&0 \\ 0&3&0&0&2&0&1 \end{bmatrix}, \quad d = \begin{bmatrix} 1&0&0&0&0&0&0 \\ 0&0&0&3&0&0&0 \\ 0&0&0&0&2&0&0 \\ 0&3&0&0&0&0&0 \\ 0&0&2&0&0&0&0 \\ 0&0&0&0&0&1&0 \\ 0&0&0&0&0&0&1 \end{bmatrix}.$$

It can be shown that H = <a,b,c,d> is isomorphic to $G_2(5)$. A presentation for H in terms of these generators is known. The group G can be defined as the group generated by elements a, b, c, d and z, where all the defining relations for H are satisfied and in addition

$$z^{-1}c_1z = x_{162},$$
$$z^{-1}c_2z = x_{163},$$
$$z^{-1}c_3z = x_{164},$$
$$z^{-1}c_4z = x_{165},$$
$$z^{-1}c_5z = x_{166},$$
$$z^{-1}a_1z = x_{167},$$
$$z^{-1}a_2z = x_{168},$$
$$z^{-1}a_3z = x_{169},$$
$$z^2 = u,$$
$$(a_4z)^3u = 1,$$
$$x_{47}zx_{364}zx_{368}z = 1,$$
$$z^{-1}x_{24}z^{-1}x_{369}zx_{23}zx_{29} = 1,$$
$$zx_{366}zx_{365}z^{-1}x_{367}z^{-1}x_{370} = 1,$$
$$zx_{96}zx_{25}z^{-1}x_{129}z^{-1}x_{371} = 1,$$

where, writing 1, ..., 8 for a, b, c, d, d^{-1}, c^{-1}, b^{-1}, a^{-1},

$$
\begin{aligned}
a_1 &= 1,\\
a_2 &= 217,\\
a_3 &= 22177,\\
a_4 &= 77122,\\
c_1 &= 3,\\
c_2 &= 831,\\
c_3 &= 732,\\
c_4 &= 237,\\
c_5 &= 68613831,\\
u &= 11,\\
x_{23} &= 4,\\
x_{24} &= 5,\\
x_{25} &= 2287822613864622832,\\
x_{29} &= 12822822613237437774323,\\
x_{47} &= 65868265,\\
x_{96} &= 24,\\
x_{129} &= 57,\\
x_{162} &= 8782671621\\
x_{163} &= 77138232\\
x_{164} &= 8783121,\\
x_{165} &= 7823712,\\
x_{166} &= 22873213776,\\
x_{167} &= 888,\\
x_{168} &= 28788,\\
x_{169} &= 1221778,\\
x_{364} &= 84768265267861171217,\\
x_{365} &= 17783262866431383875,\\
x_{366} &= 21787787322324377774323,
\end{aligned}
$$

x_{367} = 676522657677621221287,

x_{368} = 8651671658613776 7877,

x_{369} = 865287652876717717,

x_{370} = 513238651783877828,

x_{371} = 6832653831378782.

REFERENCES

[1] R. Lyons, Evidence for a new finite simple group, to appear in J. Alg.

[2] C. C. Sims, The simple group predicted by Lyons, to appear.

Rutgers University

ISOMETRIES IN FINITE GROUPS OF LIE TYPE

Bhama Srinivasan

Let $\tilde{G}$ be a connected semisimple linear algebraic group and σ a surjective endomorphism of $\tilde{G}$ such that the group of fixed points $\tilde{G}_\sigma = G$ is finite (see [1], E-4). Then G is a finite group of Lie type. Let $\tilde{T}$ be a maximal torus of $\tilde{G}$ fixed by σ and $\tilde{T}_\sigma = T$, $H = N_G(T)$, $H/T = W$. Then T is an abelian subgroup of G. We call any subgroup of G arising in this way a torus of G. We say an irreducible character of T or an element of T is regular if it is not fixed by any non-identity element of W. It has been conjectured (see [1], C-21) that there is a family of irreducible characters of G parametrized by the orbits under W of the regular irreducible characters of T.

If $\tilde{G}$ is quasi-simple there is a positive real number q associated with G; for example if $\tilde{G}$ is defined and split over a finite field and σ is the Frobenius endomorphism then G is simply a finite Chevalley group and q is the number of elements in the field ([1], E-21). Let r be the rank of $\tilde{G}$. Using the theory of exceptional characters as originally formulated by Brauer and Suzuki, T. A. Springer ([4], 7.7) has proved the following theorem which says that the conjecture mentioned above is true for large q.

THEOREM. *Suppose the non-regular elements of* T *are contained in a subgroup of order* $O(q^{r-1})$ *(i.e. a subgroup of order* $\leq kq^{r-1}$ *where* k *is a constant depending only on certain invariants associated with* $\tilde{G}$ *like its root system and not on* q*; for details see* [4]*). Then there exist* $q^r + O(q^{r-1})$ *characters of* T *for which there exist*

$q^r + O(q^{r-1})$ _characters of_ T _for which there exist corresponding irreducible characters of_ G.

This theorem is proved by seeing that the regular elements in T form a TI-set. Thus induction is an isometry from the space of complex-valued class functions on H having support on the set of regular elements in T into the space of class functions on G, and the methods of Brauer-Suzuki can be used.

Ideally one would hope to construct an isometry τ from the space of class functions on H having support on T - Z, where Z is the center of G, such that τ takes generalized characters to generalized characters. Of course in this case τ cannot be induction as T - Z will in general include non-regular elements. Now if $x \in T$, the structure of the centralizer of x in $\tilde{G}$ is known; its connected component is a reductive group ([1], E-35). Thus $C_G(x)$ can be essentially (i.e. up to extension by a subgroup of W) written as T·K where K is a normal subgroup which is a central product of groups of Lie type. A similar result is true for the subgroups $C_G(S)$ where S is a non-empty subset of T. In a theorem which will be stated in the next section we show that if isometries satisfying certain conditions can be constructed in the case of the subgroups of G of the form $C_G(S)$, $S \subseteq T$, then they can be patched together to construct an isometry from a suitable space of class functions on H into the space of class functions on G.

NOTATION. If Q is a subgroup of G, X(Q) is the space of all complex valued class functions on Q with the usual inner product. If $f \in X(Q)$, f^G is the induced class function on G. If $Q' = g^{-1}Qg$, $g \in G$, and $f \in X(Q)$, $f^g \in X(Q')$ is defined by $f^g(g^{-1}\,x\,g) = f(x)$.

If $x \in G$, $x = x_s x_u$ is the decomposition of x into its semi-simple and unipotent parts. (If G is a finite Chevalley group of characteristic p then x_s and x_u are the p'-part and the p-part of x, respectively.

V is a fixed subgroup of T, normal in H. If Q is a subgroup of H containing T, $\tilde{X}(Q)$ denotes the subspace of X(Q) of all class functions having support on T - V.

THEOREM. *Suppose we have a subset* K *of* T *such that for every non-empty subset* S *of* K, *we have an isometry*

$$\tau(S): \tilde{X}(C_H(S)) \to X(C_G(S)), \text{ satisfying}$$

(i) *If* $S' = h^{-1}Sh$, $h \in H$, *then for* $f \in \tilde{X}(C_H(S))$,

$(f^{\tau(S)})^h = (f^h)^{\tau(S')}$

(ii) *Let* $f \in \tilde{X}(C_H(S))$. *Then* $f^{\tau(S)}(x) = 0$ *if* x_s *is not conjugate in* $C_G(S)$ *to any element of* K. *Otherwise*,

$$f^{\tau(S)}(x) = \sum \frac{1}{|C_H(S) : C_H(S) \cap C_H(t)|} \times (f|C_H(S) \cap C_H(t))^{\tau(C_G(S) \cap C_G(t))}(g^{-1}xg)$$

where the sum is over all $t \in K$ *conjugate to* x_s *in* $C_G(S)$, *and* $g \in C_G(S)$ *is some element such that* $g^{-1}x_sg = t$, *so that* $g^{-1}xg \in C_G(t)$.

(iii) *For each* S, $\tau(S)$ *takes generalized characters to generalized characters*.

Then define τ: $\tilde{X}(H) \to X(G)$ *by*:

If $f \in \tilde{X}(H)$, $f^{\tau}(x) = 0$ *if* x_s *is not conjugate to any element in* K, *and otherwise*,

$$f^{\tau}(x) = \sum \frac{1}{|H : C_H(t)|}(f|C_H(t))^{\tau(C_G(t))}(g^{-1}xg),$$

where the sum is over all $t \in K$ *conjugate to* x_s *in* G, *and* $g \in G$ *is chosen as in* (ii) *above*.

Then τ *is an isometry from* $\tilde{X}(H)$ *into* $X(G)$ *and takes generalized characters to generalized characters.*

Of course the crucial thing is to prove that τ takes generalized characters to generalized characters. For this we prove the following lemma which is a variant of a result of Dade ([3], p. 172). Dade's result cannot be applied here without modificiation as his hypotheses (e.g. that H controls fusion in T with respect to G) do not hold here.

LEMMA. *For* $f \in \tilde{X}(H)$, $-f^{\tau} = \sum_{S \subseteq K} \frac{(-1)^{|S|}}{|H : N_H(S)|} (\phi^{\tau(S)})^G$,

where ϕ *is a suitable element of* $X(C_H(S))$.

In order to apply the theorem to prove the existence of irreducible characters of G corresponding to regular characters of T we have to make a suitable choice of K and V. Ideally one would hope to take V = Z and K = T - Z where Z is the center of G. If we know that the conjecture is true for subgroup $C_G(S)$ where $S \subseteq K$, and if there are enough regular characters of T, we can take generalized characters of H having support on K and having norm 2, apply τ as defined in the theorem, and obtain irreducible characters of G in the usual way. For example, let $G = G_2(q)$, $q \not\equiv 0 \pmod 6$. Then the centralizers of all the semisimple elements are known [2] and the conjecture is true for these groups. Thus the theorem can be used to show the existence of a family of irreducible characters of G corresponding to the torus of order $(q + 1)^2$. Here the tori of orders $q^2 + q + 1$ and $q^2 - q + 1$ are easier to handle as every non-identity element in these tori is regular.

In order to apply the theorem to the infinite families of

classical groups with as few restrictions as possible on the field it seems to be important to verify the conjecture for the groups GU(n,q) as they occur often as centralizers of semisimple elements in classical groups.

REFERENCES

[1] A. Borel, et. al. Seminar on algebraic groups and related finite groups. Lecture Notes in Mathematics No. 131, Springer-Verlag, 1970.

[2] B. Chang, The conjugate classes of Chevalley groups of type (G_2). J. Algebra 9 (1968), 190-210.

[3] W. Feit, Characters of finite groups. Benjamin, New York, 1967.

[4] T. A. Springer, On the characters of certain finite groups, preprint, Mathematisch Instituit der Rijksuniversiteit Utrecht, 1970.

Clark University

CENTRALIZERS OF INVOLUTIONS AND THE CLASSIFICATION PROBLEM

John H. Walter

0. INTRODUCTION

The object of this work is to present a reduction of the structure of a wide class of simple finite groups. We shall divide the class of finite groups into several subclasses, one of which we shall discuss entirely in terms of its defining property. The remaining classes fit closely certain classes of known simple groups. More will be assumed about these groups since it is reasonable to expect their characterizations. The properties we present can form the basis of an inductive approach, which will be discussed at the end. At this stage of the development of the theory our attention will ultimately be directed towards a characterization of the Chevalley groups over a field of odd characteristic.

We shall deal mainly with conditions placed on the components of the centralizers of involutions. A component L of a group H is a minimal perfect subnormal subgroup of $C_H(O_{2',2}(H)/O_{2'}(H))$. Then L/O(L) is quasisimple--that is, a perfect central extension of a simple group. Components were introduced in [4]. This definition coincides with that given there.

1. REGULATION AND BALANCED GROUPS

Our attention first is directed at generalizing the concept of balance introduced in [5]. We need some notation. Let T be the class of all 2-groups. Let $\mathcal{E}$ be the class of all elementary 2-groups. Let $\mathcal{E}_k$ be the class of all elementary 2-groups of rank at most k. When $\mathcal{X}$ is a class of groups denote by $\mathcal{X}(G)$ the set of subgroups of a group G belonging to the class $\mathcal{X}$. Let $\mathcal{E}^*(G)$ be the set of elementary 2-subgroups of G of maximum possible rank. Then if $\mathcal{E}^*(G) = \mathcal{E}_k(G)$, we say that G has 2-rank $m_2(G) = k$. The elements of $\mathcal{E}^*(G)$ play a critical role in our analysis.

Let L be a component of a group G and set $N = N_G(L)$, $C = C_G(L)$ and $\bar{N} = N/C$. A component L of a group G is said to be <u>strongly regulated</u> [<u>regulated</u>] in G provided for any $X \in \mathcal{T}(N)$ [$X \in \mathcal{E}(N)$] $O(C_N(X))\ \_\ CO(N)$. If a component is not strongly regulated, it will be said to be <u>weakly regulated</u>. A group G is said to be <u>strongly balanced</u> [<u>regularly balanced</u>] provided

(1.1) $$O(C_G(X)) \quad C_G(t)\ \_\ O(C_G(t))$$

for all $X \in \mathcal{T}(G)$ [$X \in \mathcal{E}(G)$] and all involutions $t \in C_G(X)$. This generalizes the concept of balance introduced in [5]. Strong balance is partially an inductive concept--in particular, it carries over to factor groups over central subgroups.

PROPOSITION 1.1. *A group G is strongly balanced [regularly balanced] provided that the components of the centralizer of every involution are strongly regulated [regulated] in that centralizer.*

We are ultimately interested in groups with weakly regulated components. But first we need to deal with strongly balanced groups. We set for $S \in \mathcal{T}(G)$

(1.2) $$\Gamma_{k,S}(G) = \langle N_G(R) \mid R \subseteq S \text{ and } m(R) \geq k \rangle.$$

When S is an S_2-subgroup of G, $\Gamma_{k,S}(G)$ is called the k-<u>generated core.</u> Groups different from their 1-generated core are characterized by Bender's celebrated theorem [2]. Using that result and a recent result of Shult [7], we have the following theorem.

THEOREM 1.2. *Let G be a strongly balanced group which is distinct from its 2-generated core. Then G/O(G) is isomorphic to one of the following:*

(i) A group of the form L_1L_2 where L_1 has the 2-rank at most 1 and L_2 is a normal subgroup which is isomorphic to a central extension of a group of type $L_2(q)$, $Sz(q)$ or $U_3(q)$, $q = 2^n$, $n > 1$.

(ii) The orthogonal group $O^+(4,5)$. (This contains the central product of two copies of SL(2,5) as a subgroup of index 2)

(iii) Janko's simple group of order 175,560.

Probably the above result holds without the hypothesis of strong balance provided one includes the central exeension $\hat{A}_9$ of the alternating group A_9. (Subsequent to this talk, Michael Aschbacher has obtained this result.)

We can provide one more result which yields a property which is essential in our discussion.

THEOREM 1.3. *Let G be a strongly balanced group. Let S be an S_2-subgroup. Then*

$$\langle \mathcal{V}_G(S;2')\rangle = O(G) \tag{1.3}$$

or G/O(G) is isomorphic to SL(3,3).

Because a strongly balanced group is balanced, the following result from [5] should be mentioned.

THEOREM 1.4. *Let G be a strongly balanced group with $SCN_3(2)$ nonempty. Then $O(C_G(t)) \subseteq O(G)$ for all involutions t.*

2. REGULATING SUBGROUPS

We call the elements of $\mathcal{E}^*(G)$ the *regulating subgroups* of G. They play an important role in studying weakly regulated groups. For example, if G is a Chevalley group over a field of odd characteristic whose (Weyl) rank is at least 3, then such a subgroup always contains the maximal elementary 2-subgroup of a Cartan subgroup. A regulating subgroup A acts on the set of components of G by conjugation. A product of components in an A-orbit will be called an A-component of H.

PROPOSITION 2.1. *Let* A *be a regulating subgroup of a group* H. *Assume* $m_2(Z(L/O(L))) \leq 1$ *for any component* L. *Then the* A-*components of* H *are either components or products of two components with generalized quaternion* S_2-*subgroups*.

3. SIGNALIZER FUNCTORS

Let $E \in \mathcal{E}(G)$. Set

$$\Delta_G(E) = \bigcap_{t \in E^\#} O(C_G(t)). \tag{3.1}$$

A group H is said to be k-*regulated in* H *with respect to an elementary* 2-*subgroup* A *of* H *provided* $\Delta_H(E) \subseteq O(H)$ for $E \in \mathcal{E}_k(A)$. A group G is said to be k-*balanced with respect to* A provided

$$\Delta_G(E) \cap C_G(t) \subseteq O(C_G(t)) \tag{3.2}$$

for all $E \in \mathcal{E}_k(A)$ and $t \in A^\#$.

PROPOSITION 3.1. *A necessary and sufficient condition for a group* G *to be* k-*balanced with respect to a regulating subgroup* A *is that the* A-*components* L *of the centralizers* H *of the involutions of* A be k-regulated in $N_H(L)$.

This imposes a restriction on the 2-rank for the groups we consider. Except for the groups PSL(2,q), the Chevalley groups over fields of odd characteristic are 2-regulated. The group PSL(2,q) always appears as a 3-regulated compoeent. The groups $\hat{A}_{2^r+3}$ are awkward; one must take k = 4 when these components are present.

In [6], the following theorem has been established.

THEOREM 3.2. (Gorenstein) *Let* G *be a* k-*balanced group with respect to a subgroup* $A \in E(G)$. *Suppose that* $m(A) \geq k + 2 \geq 4$. *Then* $\langle \Delta_G(T) | T \in E_k(A) \rangle$ *has odd order*.

We shall introduce the groups

(3.4) $$W_A = \langle \Delta_G(T) | T \in E_k(A) \rangle$$

(3.5) $$M_A = N_G(W_A).$$

COROLLARY 3.3. $M_A \supseteq \Gamma_{k+1,A}(G)$.

4. ADMISSIBLE COMPONENTS

We now prescribe conditions on the weakly regulated components of a group H. We first need some definitions and notation. We define a *solvable component* to be a minimal subnormal subgroup L such that L/O(L) is isomorphic to PSL(2,3) of SL(2,3).

Let $L(H)$ be the set of components of a group H. Let L_2 be the set of strongly regulated components. Let L_1 be the set of regulated but not strongly regulated components. Let L_0 be the set of weakly regulated but not regulated components. Thus for any finite group $L = L_0 \cup L_1 \cup L_2$.

Denote by $\Lambda(H) = \langle L | L \in \mathcal{L} \rangle$. Then $\Lambda(H)/O(\Lambda(H))$ is a central product of quasisimple groups $LO(\Lambda(H))/O(\Lambda(H))$. A central product of quasisimple groups will be called <u>semisimple</u>. Denote

$$\Lambda_i(H) = \langle L | L \in \mathcal{L}_i \rangle,\ i = 0,1,2.$$

$$\Lambda_i'(H) = \Lambda_i(H)O(H),\ i = 0,1,2.$$

$$\Lambda_s(H) = \langle L | L \text{ is a solvable component of } H \rangle.$$

Let L be a weakly regulated component of a group H. Set $\overline{L} = L/O(L)$. We say that L is <u>admissible</u> in H provided the following conditions hold.

I. For each involution $t \in N_H(L)$, $m_2(Z(C_{\overline{L}/Z(\overline{L})}(t))) \leq 1$, and $O_{2',2}(C_L(t))\Lambda(C_L(t)) \supset O_{2',2}(C_L(t))$.

II. For each $T \in \mathcal{E}_2(N_H(L))$, one of the following holds

(i) $O(C_L(t)) \neq 1$ for some $t \in T^\#$ and $L = \langle \Lambda_0'(C_L(t)) \mid t \in T^\# \rangle$

(ii) $O(C_{\overline{L}}(t)) = 1$ for all $t \in T^\#$

(iii) $O(C_{\overline{L}}(t)) = O(C_{\overline{L}}(T))$ for all $t \in T^\#$ and

(a) $|O(C_{\overline{L}}(T))| = 3$

(b) $O(C_{\overline{N}}(t)) = O(C_L(t))$ where $N = N_H(L)$ and $\overline{N} = N/O(N)$.

(c) The representations of L over fields of characteristic 3 are 3-stable.

III. For each $T \in \mathcal{E}_3(N_H(L))$, either L is contained in $\Gamma_{2,T}(H)$

or $\bar{L}$ is isomorphic to the central extension $\hat{A}_9$ of the alternating group A_9.

In (II), condition (i) forces L to be not regulated. The prototypes for the groups specified in (iii) are the alternating groups. Components satisfying the first two conditions of (II) above will be called <u>regular</u>; components satisfying the last condition will be called <u>irregular</u>.

Now let $A \in E^*(G)$ for some group G. Suppose $m_2(G) > k$ for some integer k. We say that G has k-<u>admissible</u> weakly regulated A-components provided the weakly regulated components of the centralizers of involutions of A satisfy the following conditions.

IV. For each involution $t \in A$ and each weakly regulated component L of $C_G(t)$, set $\bar{L} = L/O(L)$ and stipulate

(i) $\Lambda(C_{\bar{L}}(t))$ is semisimple and

$$O_{2',2}(\Lambda_s(C_{\bar{L}}(t)) = O_2(\Lambda_s(C_{\bar{L}}(t))$$

(ii) Let $T \in E_k(A)$; then

$$L = \langle \Lambda_1'(C_L(t))O_{2',2}(\Lambda_s(C_L(t))) \mid t \in T^\#\rangle. \tag{4.1}$$

(iii) L is k-balanced with respect to A.

(iv) $L \subseteq \Gamma_{k+1,A}(C_G(t))$.

5. STRUCTURE OF CENTRALIZERS OF INVOLUTIONS BELONGING TO A REGULATING 2-SUBGROUP

THEOREM 5.1. *Let G be a finite quasisimple group, and let $A \in E^*(G)$. Assume that the weakly regulated components of the centralizers of involutions and of proper local subgroups of G of maximal possible 2-rank are admissible. Assume that G has* k-admissible weakly regulated A-components. Assume $m_2(G) \geq 5$.

Then $W_A = O(G)$ and the A-components of G are quasisimple.

In applying this theorem, k should be taken to be 3 or 4 with the higher value chosen in case groups of the form A_{2^r+3} are allowed as components. In this case it will also be necessary to have $m(A) \geq 7$ in order to verify (IV-iv). Eliminating these groups and just allowing Chevalley groups over fields of odd characteristic as weakly regulated components would a low us to take $m(A) \geq 5$.

It is anticipated that Theorem 5.1 will be employed in an inductive argument characterizing the class of regulated semisimple groups. Thus we anticipate that groups of low 2-rank will be characterized by other means. The work of Gorenstein and Harada represents a substantial step in this direction.

By concentrating our attention on the Chevalley groups over fields of odd characteristic, we can go further. We say that a weakly regulated component L of a group H is L_0-generated provided

$$(5.1) \qquad L = \langle \Lambda_0'(C_L(t)) \quad t \in T\# \rangle$$

for any $T \in \mathcal{E}_2(N_H(L))$. We say that an A-component L is of type $\mathcal{C}_0$ provided $\Lambda_0'(C_L(t))$ has at most two A-components for any involution $t \in A$. We say that an A-component L of G is standard provided $C_G(L)$ has generalized quaternion S_2-subgroups.

THEOREM 5.2. Let G be a finite simple group and let $A \in \mathcal{E}^*(G)$. Assume that G satisfies the hypothesis of Theorem 5.1 with respect to A. Assume that the A-components of G are L_0-generated and type $\mathcal{C}_0$. Then some A-component of G is standard; and for every involution $t \in A$, $\Lambda(C_G(t)) = \Lambda_0(C_G(t))$ has at most two A-components and $m_2(Z(\Lambda(C_G(t))) = 1$.

The above result and its proof suggest that it might be possible to show that G itself satisfies the conditions of a weakly regulated admissible component. Unfortunately the argument will not work if G is a quasisimple group with nontrivial center, and this case must be considered in order to deal inductively with the class of groups considered in Theorem 5.1. The assumptions on the A-components in the above theorem can be used in Theorem 5.1 to replace the assumption that the regulated components of proper subgroups of maximal 2-rank are admissible.

Assuming G is a simple group of 2-rank at least 10 whose A-components are Chevalley groups of type A_n, G itself is a Chevalley group of type A_n. This follows from the thesis of Patrick Cassady [3].

REFERENCES

[1] H. Bender, Doubly transitive groups with no involution fixing two points. (preprint; see [2], p. 130 this vol.)

[2] __________, Finite groups having a strongly embedded subgroup. (preprint; see [2], p. 130 this vol.)

[3] P. Cassady, thesis, University of Illinois, 1972.

[4] D. Gorenstein and J. H. Walter, The π-layer of a finite group, Illinois J. Math <u>15</u>

[5] __________, Centralizers of involutions in balanced groups, J. Algebra (to appear)

[6] __________, Centralizers of involutions in Λ-groups, (to appear)

[7] E. Shult, On the fusion of an involution in its centralizer (to appear)

University of Illinois

A CHARACTERIZATION OF ORTHOGONAL SIMPLE GROUPS $P\Omega(2n,q)$

W. J. Wong

Suppose q is a prime power, $q \equiv \varepsilon \pmod 4$, $\varepsilon = \pm 1$. Let $O_\alpha(2n,q)$ denote the orthogonal group of a non-degenerate symmetric bilinear form of discriminant α on a 2n-dimensional vector space over the field F of q elements. (There are essentially 2 cases, depending on whether or not α is a square in F.) Let $O^+_\alpha(2n,q)$ be the subgroup of $O_\alpha(2n,q)$ consisting of elements of determinant 1, $\Omega_\alpha(2n,q)$ its commutator subgroup, and $P\Omega_\alpha(2n,q)$ the corresponding projective group. For $n \geq 3$, the last group is simple, and we are concerned with characterizing it among all simple groups by the structure of the centralizer of an involution. By examining the centralizer of an involution in $P\Omega_\alpha(2n,q)$ corresponding to an involution of $\Omega_\alpha(2n,q)$ which fixes the vectors of a (2n-2)-dimensional subspace, we are led to the following hypotheses.

(1) G _is a finite group_, _containing an involution_ t _whose centralizer_ C(t) _has a normal subgroup_ M _isomorphic to_ $\Omega_\alpha(2n-2,q)$.

(2) $|C(t) \cap C(M)| < \frac{3}{2}(q - \varepsilon)$.

(3) C(t) _has two involutions_ u, u', _which induce the automorphisms of_ M _which correspond to automorphisms of_ $\Omega_\alpha(2n-2,q)$ _induced by involutions_ z, a' _of_ $O_\alpha(2n-2,q)$, _one from each of the cosets of_ $\Omega_\alpha(2n-2,q)$ _which do not lie in_ $O^+_\alpha(2n-2,q)$.

(4) $|C(t) : \langle M,u,u'\rangle|$ _is odd_.

(5) $[u,u']h^{-1}$ _has even order, where_ h _is the element of_ M _corresponding to the element_ [z,z'] of $\Omega_\alpha(2n-2,q)$.

THEOREM. *If* G *satisfies the hypotheses* (1) - (5) *and* n *is large enough, then either* $G = C(t)o(G)$ *or* G *is a semidirect product*

$$G = AG_0, \quad G_0 \triangleleft G, \quad A \cap G_0 = 1,$$

where G_0 *is isomorphic with* $P\Omega_\alpha(2n,q)$ *or* $\Omega_\alpha(2n,q)$, A *is cyclic of odd order, and* A *acts faithfully on* G_0 *by field automorphisms*.

For $\varepsilon = 1$, $n \geq 5$ is large enough, and for $\varepsilon = -1$, $n \geq 8$ is large enough. These inequalities may be weakened slightly in some cases, depending on whether or not α is a square and whether or not n is even. In its present form, the theorem cannot hold for $n = 4$ when α is a square, because of the existence of the triality automorphism of $P\Omega_\alpha(8,q)$. We remark that for $n = 3$, $P\Omega_\alpha(6,q)$ is isomorphic with PSL(4,q) or PSU(4,q). These groups have been characterized by the structure of the centralizer of an involution by Phan [1] and [2].

In outline, the proof of the theorem resembles the argument used in characterizing the symplectic simple groups PSp(2n,q) [3], with complications due to the fact that there are two non-degenerate symmetric bilinear forms in each dimension. A study is made of the possible fusion patterns for the involutions of C(t). If $\varepsilon = 1$, this provides elements corresponding to elements of the Weyl group of $\Omega_\alpha(2n,q)$ not in the Weyl group of $\Omega_\alpha(2n-2,q)$, and the construction of the subgroup G_0 is then straightforward. If $\varepsilon = -1$, an additional argument is needed to obtain the missing elements of the Weyl group. The necessity of having enough room to carry out this argument is the explanation for the exclusion of low values of n in this case.

REFERENCES

[1] K. W. Phan, A characterization of four-dimensional unimodular groups, J. Alg. 15 (1970), 252-279.

[2] __________, A characterization of the unitary groups $PSU(4,a^2)$, q odd. J. Algebra 17 (1971), 132-148.

[3] W. J. Wong, Characterization of the finite simple group $PSp_{2n}(q)$. J. Algebra 14 (1970), 531-551.

University of Notre Dame